T. D. (Thomas Davison) Crothers

The Disease of Inebriety from Alcohol, Opium, and Other Narcotic

Drugs

It's Etiology, Pathology, Treatment and Medico-Legal Relations

T. D. (Thomas Davison) Crothers

The Disease of Inebriety from Alcohol, Opium, and Other Narcotic Drugs
It's Etiology, Pathology, Treatment and Medico-Legal Relations

ISBN/EAN: 9783744669450

Printed in Europe, USA, Canada, Australia, Japan

Cover: Foto ©berggeist007 / pixelio.de

More available books at **www.hansebooks.com**

THE

DISEASE OF INEBRIETY

FROM

ALCOHOL, OPIUM AND OTHER NARCOTIC DRUGS,

ITS ETIOLOGY, PATHOLOGY, TREATMENT
AND
MEDICO-LEGAL RELATIONS.

*ARRANGED AND COMPILED BY THE AMERICAN
ASSOCIATION FOR THE STUDY AND
CURE OF INEBRIETY.*

NEW YORK:
E. B. TREAT, PUBLISHER,
No. 5 Cooper Union.
1893.   [PRICE, $2.75.]

DEDICATED

TO THE MEMORY OF

DR. J. EDWARD TURNER,

THE PROJECTOR AND FOUNDER OF THE FIRST
INEBRIATE ASYLUM IN THE WORLD, AT
BINGHAMPTON, NEW YORK.

ALSO TO THE MEMORY
OF THOSE PIONEERS, NOW PASSED AWAY, WHO
FAR IN ADVANCE OF THEIR TIMES
ORGANIZED THE FIRST SCIEN-
TIFIC SOCIETY FOR THE
STUDY AND CURE OF
INEBRIETY :

DRS. JOSEPH PARRISH, THEODORE L. MASON, WIL-
LARD PARKER, DANIEL DODGE, JOHN WILLETTS,

AND OTHERS.

THIS VOLUME IS OFFERED AS A SLIGHT TRIBUTE
TO THEIR SUPERIOR DISCERNMENT, AND PRAC-
TICAL RECOGNITION OF THE GREAT ONCOM-
ING TRUTHS OF INEBRIETY.

INTRODUCTION.

In 1870 the American Association for the Study and cure of Inebriety was organized in New York City. Its members were composed of physicians connected with asylums for Inebriates and others interested in the scientific study of the drink problem.

The following statements were adopted as representing the principles and purposes of the Association :

1. Inebriety is a disease.
2. It is curable as other diseases are.
3. The constitutional tendency to this disease may be either inherited or acquired ; but the disease is often induced by the habitual use of alcohol or other narcotic substances.
4. Alcohol has its place in the arts and sciences, but as a medicine it is classed among the poisons, and its internal use is always more or less dangerous, and should be prescribed with great caution.
5. All methods hitherto employed for the treatment of inebriety that have not recognized the disordered physical condition caused by alcohol, opium, or other narcotics, have proved inadequate in its cure ; hence the establishment of HOSPITALS for the special treatment of inebriety, in which such conditions are recognized, becomes a positive need of the age.
6. In view of these facts, and the increased success of the

treatment in inebriate asylums, this Association urges that
every large city should have its local and temporary hospital
for both the reception and care of inebriates; and that
every State should have hospitals for their more permanent
detention and treatment.

7. Facts and experience indicate clearly that it is the
duty of the civil authorities to recognize inebriety as a disease, and to provide means in hospitals and asylums for its
scientific treatment, in place of the penal methods by fines
and imprisonment hitherto in use, with all its attendant evils.

8. Finally, the officers of such hospitals and asylums
should have ample legal power of control over their patients,
and authority to retain them a sufficient length of time for
their permanent cure.

For twenty-two years this association has held annual
and semi-annual meetings in which a large number of
papers have been presented, read and discussed.

In the first seven years, six volumes of transactions
were issued, then a society journal was established as the
official organ of the association. This was called the
Quarterly Journal of Inebriety, which has been published
since 1877 ; the special work of this journal has been to
gather and group the scientific literature of this subject
and make it available for future study.

In addition to this literature, many members of this
association have published volumes on this subject. Of
these may be mentioned : " *Alcoholic Inebriety*," by the late
Dr. Joseph Parrish, President of the Association ; " *Metho-
mania*," by Dr. Albert Day, the present President ; " *Inebriism*," by Dr. T. L. Wright, a member of the Association ;
" *Inebriety*," by Dr. Norman Kerr, of London, an honorary
member ; " *Alcoholism*," by M. Magnan, an honorary member,
of France ; " *The Manifestations of Alcoholism*," by Dr. Lentz,
of Brussels, an honorary member ; " *Alcoholism as a Malady*,"
by Dr. A. Baer, of Berlin, also an honorary member.

In both this country and Europe, a large number of valuable papers and monographs on this topic, have appeared.

Among the most prominent who have written on this subject may be mentioned Drs. Mason, Crothers, Mattison, Mann, Hughes, Searcy, Wright, Davis, Shepard, Morris, Day and many others. Abroad, Drs. Peddie, Kerr, Clouston, Carpenter, Ridge, Richardson and many of the most eminent physicians and medical teachers of Europe.

A large part of these writings have been published in the *Journal of Inebriety*, together with the discussions and criticisms which have grown out of them.

This literature has been copied into medical and other journals and become the subject of lectures, sermons and works on temperance ; but in many cases the original ideas have been changed and modified to such an extent as to be unrecognized. As a result, the most confusing uncertainty and conflict of opinion exists concerning the scientific study of inebriety and its maladies.

Recently, this subject has attracted renewed attention, growing out of the empiric assumptions of specific remedies for its cure. Following this, an increasing demand has appeared for some authoritative grouping of the studies of scientific men in this field, and suggested the value of a volume comprising some of the best papers which have been published in the journal.

In accordance with this, the following resolution was unanimously adopted at the November meeting of the Association for the Study and Cure of Inebriety.

" *Whereas,* In view of the fact that the demand for copies of the *Journal of Inebriety* containing special studies, monographs and papers on distinct phases of inebriety is steadily increasing, together with constant inquiry from physicians for facts, statistics and conclusions relating to the disease of inebriety;

" *Therefore, resolved,* that the secretary be authorized to prepare a volume, which shall contain the most reliable conclusions and studies of eminent authorities on all phases of the disease of inebriety up to the present time ;"

" *Resolved,* that the secretary, with the publication committee, make all arrangements for the printing and issuing of said volume at the earliest possible moment.

This volume is in conformity with this resolution, and simply aims to give the reader many of the best and most suggestive studies of this subject which have appeared.

More prominence has been given to heredity, treatment and medico-legal relations, because these phases are disputed and the subject of confusing controversy. The fact of disease in inebriety has passed beyond question and is practically settled, hence all moral theories or discussions are of little or no value from a scientific point of view. These various studies are not presented as exhaustive, but rather as outlines suggesting wide stretches of the unknown, inviting the reader to take up this subject and extend its frontiers far in advance of the present.

It will be apparent to all that the scientific study of inebriety is yet in its infancy, and no student or volume can do more than give a general view of the most probable facts which appear to be sustained by the strongest evidence available up to this time.

This volume is intended to represent the work of the Association for the Study and Cure of Inebriety and the character of the papers and discussions which have appeared in its journal.

For over a quarter of a century, this association has studied exclusively the scientific side of inebriety, and although it has not attracted much attention and is practically unknown, it has been a great silent power, crystallizing and marshalling the many facts constantly appearing from a knowledge of these cases.

The drink malady, its causes, cure and prevention, are practically an unoccupied field of medicine, and this work is offered as a guide and stimulus to farther study in this direction.

These selections have been gathered from over five thousand pages of printed matter, published in the journal and transactions, and are from papers which have not appeared elsewhere and hence will be new to most physicians. It will be evident to the reader that while the facts are very numerous and startling, and fully sustain the principles of the association, they are not yet sufficiently studied and generalized to be accepted as absolute truths.

This volume brings the subject up to the frontier lines, and suggests that the possibility of restoring the inebriate and stamping out inebriety, is only limited by our want of knowledge of the laws and means to accomplish this end. It also suggests that a solution of this great problem of the forces and sources which develop the armies of the inebriates or halt and disband them, must be along the line of accurately observed facts and their meaning.

The purpose of this work will be accomplished, if, in any way, it shall help on this pioneer work.

T. D. CROTHERS,
Secretary.

HARTFORD, CONN.
March, 1893.

CONTENTS.

CHAPTER I.

CHAPTER VIII.

CHAPTER XIX.

CHAPTER XX.

..... CHAPTER XXI

CHAPTER XXII.

CHAPTER XXIII.

CHAPTER XXIV.

CHAPTER XXV.

CHAPTER XXVI.

CHAPTER XXVII.

CHAPTER XXVIII.

CHAPTER XXIX.

CHAPTER XXX.

CHAPTER XXXI.

CHAPTER XXXII.

CHAPTER XXXIII.

CHAPTER XXXIV.

CHAPTER XXXV.

CHAPTER XXXVI.

CHAPTER XXXVII

CHAPTER XXXVIII.

DISEASES OF INEBRIETY.

CHAPTER I.

EARLY HISTORY OF DISEASE THEORY.

It is a curious fact that inebriety was recognized as a disease long before insanity was thought to be other than spiritual madness and a possession of the devil. The fact has escaped the attention of persons who assert that inebriety is always a vice, and the disease theory is only an extravagant view of enthusiasts, peculiar to our times. For over a century, the disease of insanity was denied and contested. Inebriety is passing the same ordeal of ignorant opposition and criticism, notwithstanding it has been recognized by a majority of the leading physicians of the world to-day.

The following outlines of the early history of inebriety will show that this is not a mere theory of our times, but a great truth, outlined and foreshadowed by many of the leading physicians and philosophers of the past.

This disease was hinted at in an early age of the world, and is by no means a modern idea. On an old papyrus found in one of the tombs of Egypt, dating back to a very ancient period, was the following significant passage, referring to an inebriate who had failed to keep sober :.

17

" Thou art like an oar started from its place, which is
unmanageable every way ; thou art like a shrine without its
God, like a house without provisions, whose walls are found
shaky."

Many of the sculptures of Thebes and Egypt exhibit
inebriates in the act of receiving physical treatment from
their slaves, such as purgatives, rubbings or applications to
the head and spine.   Heroditus, five centuries before the
Christian era, wrote, " that drunkenness showed that both
the body and soul were sick."   Diodorus and Plutarch
assert " that drink madness is an affection of the body
which hath destroyed many kings and noble people."
Many of the Greek philosophers recognized the physical
character of inebriety and the hereditary influence or tendencies which were transmitted to the next generation.

Laws were enacted forbidding women to use wine, and
young boys were restricted.

Frequent reference is made to the madness which
sought solace in wine and spirits.   Such cases are called
and urged to give more diligent care to their bodies, a
distinct hint of the physical origin of inebriety.

In the first century of the Christian era, St. John
Chrysostom urged that inebriety was a disease like dyspepsia, and illustrated his meaning by many quaint reason-
ings.   This was the first clear, distinctive recognition of
the disease, which had been hinted at long before.

In the next century, Ulpian, the Roman jurist, referred
to the irresponsible character of inebriates, and the
necessity of treating them as sick men.   His views were
embodied in some laws, which referred more distinctly to
the physical nature and treatment of inebriety.

Many of the early and later writers of Roman civilization contain references to drunkenness as a bodily disorder,
not controllable beyond a certain point, which resulted in
veritable madness.   Nothing more was heard of this theory
until the thirteenth century, when one of the Kings of

Spain enacted laws fully recognizing inebriety as a disease, lessening the punishment of crime committed when under the influence of spirits. One of the laws provided, that when murder was committed during intoxication, the death penalty should be remitted, and the prisoner be banished to some island for a period of not less than six years.

In the sixteenth century the penal codes of France and many of the German States contained enactments which recognized the disease character of inebriety.

All punishment for crime committed during this state varied according to the conditions of the prisoner at the time.

Drunkenness, continued beyond a certain point, was regarded as a condition of insanity and irresponsibility. In many of the medical writings of these ages, drunkenness and madness were mentioned as synonyms, and curious reflections on the nature and treatment of the evil are detailed.

In 1747, Condillac, a French philosopher, wrote expressing clear views of the disease of inebriety, also that the State should recognize and provide means for its treatment. He asserted that the impulse to drink was like insanity, an affection of the brain, which could not be reached by law or religion.

Dr. Benjamin Rush, of Philadelphia, in 1790, set forth the same theory, supported by a long train of reasoning. To him belongs the honor of first elaborating this subject and outlining what has been accepted half a century after.

In two essays, entitled, " The influences of physical causes upon moral faculties," and " An inquiry into the effects of ardent spirits upon the human body and mind," he described the disease of inebriety, dividing it into acute and chronic forms, giving many of the causes, of which heredity was prominent, also urging that special measures be taken in the treatment, which should be in a hospital for this pur-

pose. Up to this time these views were clear and distinctive, although not published until 1809. They were entirely
independent of all previous observations.

In 1802, Dr. Cabanis, of Paris, wrote fully indorsing the
views of Condillac, that inebriety, like insanity, was a disease
which should be studied, and that it was a distinct form of
mental disorder, needing medical care and treatment. Professor Platner, of Leipsig, published a paper in 1809 (the
same year. Rush's writings appeared), affirming that inebriety was like an insane impulse and a form of insanity
which should receive medical care and be studied by the
aid of science.

In 1817, Salvator, of Moscow, a physician of some eminence, published a pamphlet called " Ebriosity, its pathology
and treatment." He divided drunkenness into two forms,
intermittent and remittent, and urged that they be treated
by physical means.

Esquirol, in 1818, described a condition of the nervous
system in which inebriety was sure to follow. In 1822, Buhl
Cramner, a distinguished physician of Berlin, wrote a small
book defining inebriety as a state of irritation of the brain
and nervous system, to be cured by physical means because
it was purely of physical origin.

In Europe the writings of these four men, Platner, Salvator, Esquirol and Cramner, placed the subject on a scientific basis, paving the way for a wider and more thorough
study. Although Dr. Rush had written on the general
subject more definitely than the others, his writings were
practically unknown.

In 1830, a committee of the Connecticut State Medical
Society, appointed the year before to report on the necessity
of an asylum for the medical care and treatment of inebriates, recommended in an elaborate essay that it was expedient to establish an asylum for the cure of inebriates.

This report, written by Dr. Todd, the superintendent of

the insane asylum, strongly urged the recognition of the disease of inebriety and its curability by physical means.

This report, which may be found in the Transactions of the Conn. State Medical Society, was a remarkable conception of the subject, far beyond the current opinions of that time, only now beginning to be recognized.

In 1833, Dr. Woodard, of Worcester, Mass., urged that inebriety was a disease, and curable as other diseases in asylums for this purpose. This attracted attention and was endorsed by many. In 1839, Dr. Roesch, of Tubingen, in a volume on spirituous liquors, urged that inebriety be considered a disease. He elaborated the idea and pointed out the errors of other authors at some length.

Dr. Nasse, of Bonn, followed, urging the same view.

From this date, the literature has grown rapidly, especially in this country. The opinions and writings of American physicians on this subject are now widely sought for in Europe. The English Lunacy Commission, in 1844, urged that inebriates should be regarded as insane, sent to asylums for treatment, and not punished as before.

Many other writers timidly followed these advanced theories and, when stoutly opposed, compromised by admitting a state of half disease and half vice.

CHAPTER II.

The Washingtonian temperance revival, of 1840, seemed to have cleared away many of the old theories and prejudices, and gave clearer conceptions of the real nature and character of inebriety. The sudden and intense projection of the moral side of inebriety reacted when this reform wave died away, but it had served to fuse and mobilize the tide of oncoming truth, outlined for a long time in frequent statements of the disease of inebriety. The first practical demonstration of this began in a lodging-house for inebriates, opened in 1845, at Boston, Mass., called the Washingtonian Hall. This recognized, in a crude way, the physical disabilities of inebriates, and sought to remove them by physical means. This was the first embrio asylum of the world, which has grown into the Washingtonian Home of the present day.

While physicians and writers were repeating the theories of the possibility of disease of inebriety, Dr. J. Edward Turner, a Maine physician, came forward as the great pioneer to vitalize and show the practical value of the truth that inebriety was a disease, and curable.

In 1846, Dr. Turner became interested in this work from an ineffectual effort to save an early friend who was an inebriate. He recognized the nature of the disease of inebriety and the need of hospital treatment, and began an enthusiastic agitation of the subject. After eight years of most persistent effort, in the face of great opposition, he succeeded in enlisting the attention of many eminent medi-

22

cal men, and in forming a company to build an Inebriate Hospital, with the late famous surgeon, Dr. Valentine Mott, as president. Laws were passed to give power to hold inmates, a charter was granted, and nearly fifty thousand dollars subscribed for the grounds and building. Ten years later, in 1864, a magnificent building was completed and opened for patients, at Binghampton, N. Y., the pioneer hospital of the world. Later, a fire destroyed part of the building, which was soon after rebuilt. Then the board of trustees became involved in a controversy with the founder, Dr. Turner, resulting in his retirement, and placing the hospital in the care of the State. Passing into the management of politicians, its history was a series of misfortunes, until finally it was changed into a chronic insane asylum. Thus Dr. Turner, the founder, who had conceived and built this hospital, giving over a quarter of a century of time and effort, and his own personal fortune, shared the fate of all reformers and benefactors of the world, in the obloquy and disgrace of being driven away from the creation of his own genius.

The New York State Inebriate Asylum was the first institution of the kind ever organized. After fourteen years of a most extraordinary history, it was closed and converted into an insane asylum. The same old battle has been waged around and about this institution with which every new truth of science has had to contend. Bitter skeptics and foolish enthusiasts have rushed to conclusions in regard to his work, which could only be determined by a century of accurate study and observation. Literally, this pioneer asylum has been the great "On to Richmond" movement, checked by a Bull Run disaster. During the fourteen years of its existence, two thousand three hundred and forty-four inebriates were under treatment. A vast majority of these were the most chronic cases.

The enthusiasm which at first centered about Binghampton Hospital reacted, and the moralists and temperance

advocates, who from the first had opposed this movement
as an "infidel work" to diminish human responsibility,
used every means to spread the idea of failure and condemn
other efforts in this direction. But a great fact had been
recognized, and its practical character had taken deep root
in the public mind. The birth of the Binghampton Hospital
was followed by the organization of over a dozen different
hospitals for this work in different parts of the country,
some of which are still in existence, doing grand work.

Notwithstanding the misfortunes of the first hospital
and its founder, a large number of similar places have been
organized and managed with success. As in all new enterprises, many of these hospitals must suffer from non-expert
management, and be organized on some theory of the
nature and treatment of inebriety not founded in correct
study and experience. After a time they are abandoned
or changed to homes for nervous and insane cases. Over
fifty different hospitals for inebriates have been established
in America. More than thirty of this number are in successful operation ; the others have changed into insane
asylums, water-cures, etc. Three large buildings or institutions are practically "faith cures," where all physical
remedies and means are ignored. Several asylums are
called homes for nervous people, to conceal the real cause,
and thus protect the patients from the supposed stigma of
inebriety. Others are literally lodging-houses, where the
inmates can remain a few days and recover from the effects
of spirits. Several places make a specialty of opium cases ;
in some the treatment is often empirical. In only a few of
these hospitals is inebriety studied and treated on a scientific basis. The others are passing through the ordeal of
"elimination and survival of the fittest," incident to every
new advance of science. In many of the States large public hospitals are projected, and are awaiting pecuniary aid
from the State or from other sources.

Most of the inebriate hospitals in America are private

and corporate organizations, which receive, from time to time, State aid. Some of them have endowments, such as free beds, or incomes from estates, or are given so much of the license money. Others depend upon the income from patients, private donation, and charities generally. Very few paupers or indigent poor are received in any of these hospitals. This class appear in the " lodging " and " faith-cure " places. The State of Connecticut has projected a workhouse hospital for the criminal class of these cases, where the commitment is for three years, but want of State aid has prevented practical work so far. In three other States similar projects for the pauper inebriates have been organized, but for various reasons have not gone into oper-ation.

In Europe, over sixty hospitals for the physical care and treatment of inebriates are in active operation to-day. There are two in Australia, one in China, two in India, one in Ceylon, three in Africa, and one in Mexico. While these institutions vary widely in plan and methods, the central idea is the physical treatment of the inebriate and his malady.

The literature of this subject has grown to enormous proportions, principally in papers, lectures and pamphlets.

The exact scientific study of inebriety has revealed facts and conclusions which have aroused bitter controversy from those who should have been first to welcome its truths.

A large number of so-called reformers have been from the first denouncers of the disease theory, but unwittingly they have greatly helped on the work by controversy and agitation. In 1870, an association was formed of persons who were connected with inebriate asylums, or interested in the subject, called the American Association for the Study and Cure of Inebriates. This Association has since met yearly and semi-yearly ; its papers and transactions comprise the first permanent literature on this subject. In 1877, the *Journal of Inebriety* was issued as the organ of this association, and from that time has been the medium through which most of the literature of the subject from America has been presented. Inebriety, and its curability in special hospitals, has been the cardinal principle of this Association, which it has demonstrated and maintained with increasing vigor. Discussing exclusively the physical side of this subject, amid difficulties which only pioneers in a new field have to contend with, the foundation of a scientific literature has been laid which will be historic to future students of this subject. This Association and its journal are great silent powers, crystallizing and marshaling the many facts constantly appearing from a study of these cases. Already several members of this Association have published works on inebriety that mark the beginning of a " new era," and have given great impetus to its scientific study.

In 1884, a similar society was formed in England, called the Society for the study and cure of Inebriety. Similar societies have been formed in France, Switzerland, Germany and Sweden.

In 1887, an International Congress was held in London for the discussion of the disease of inebriety. Delegates from all the world were present.

Numerous committees, scientific and legislative, have made elaborative reports on the disease of inebriety and the means and measures for its prevention and treatment.

CHAPTER III.

This term more accurately represents the general phenomena called drunkenness than any other. There are two great classes, which may be designated Inebriates and Dipsomaniacs.

The first class, when studied collectively, seem to divide into several classes from different causes with different symptoms and termination. One of these classes are the *accidental idebriates,* persons whose use of spirits are dependent entirely on external conditions and environment. They seem to possess a high degree of brain and nerve instability, and have small resisting power, or capacity to adapt themselves to new surroundings and conditions of life, and retain any fixed individuality. Such persons drink or abstain according to the company they are associated with. They are enthusiastic temperance reformers, or, abject drunkards, from the mere contagion of surroundings. In a prohibition community, they are models of sobriety, or types of the lowest drinkers where spirits are free. Their progress and future turn altogether on environment.

A second class of these accidental inebriates are controlled solely by functional disturbances of the body. Spirits are never used except when some derangement of the organism occurs. Attacks of indigestion, chills, overwork, and all forms of minor strains or drains, lead up quickly to the use of spirits, without reason or premeditation. After the sedative action of alcohol has taken place a

27

few times, the internal resisting force is permanently
broken up and the person drinks ever after.

A second group may be called *emotional inebriates*. They
belong to that class who are always on the borders of
hysteria, with feeble and unstable will. Wayward, selfish,
fitful, uncertain, out of harmony with every relation of
life ; all the time suffering from a constitutional unrest,
and emotional struggles to attain the impossible. Spirits
afford relief for this physical state, and are alternately
used and abandoned until death or insanity comes on.

Another class, called *solitary inebriates*, use spirits only at
night or alone, and seem morbidly sensitive to conceal this
fact. They always deny all use of spirits, and display great
cunning and tact to cover up this condition. Their drinking seems often to be under control of the morbid reason,
and is checked at certain times, and is limited by conditions and circumstances. These are the most interesting
of all cases, and often appear in persons who occupy
responsible positions, and involve very important questions of business and science.

A large class of inebriates, appearing in lower levels of
life, are found to be literally constitutional paupers, who
through heredity and environment are switched off on the
side-track of dissolution. They may be termed the
pauper inebriates, and are persons with special unfitness for
healthy life, who are by the higher processes of nature
crowded out in the race struggle. Inebriety is simply
another symptom of this condition. The subsidence of
the drink craze is followed by other degrees of degeneration. Criminality, pauperism, prostitution, are all allied
conditions. Many other sub-classes of inebriety are found,
but they all date from some condition of brain and nerve
degeneration, and although the symptoms may vary, the
line of progress and growth are practically the same.

CHAPTER IV.

DIPSOMANIA : ITS HISTORY AND RELATION.

Dipsomania is a term used to designate a large class of inebriates, in which the drink impulse comes on suddenly and after a time dies out, and is succeeded by a free interval. This has been truly termed a neurosis, which is practically another branch of the same family of epileptics, insane hysterias and paranoic.

There are three principal forms : First, the acute, which is not very common, where the drink craze comes on suddenly from hemorrhage sunstroke ; and also convalescence from fevers, excessive strains, as continuous overwork, with mental care and excitement, and also severe drains of the nerve energy, exhaustion and other causes.

This variety seems to resemble an attack of mania in its violence and uncontrollability, and good evidence of dipsomania being an insanity and requiring careful watching and treatment.

The *periodic* is a more common variety. By some authors it is considered the most common form, but in our experience the chronic is more frequent. In this variety are classed those patients who have periodic attacks of drinking lasting for a long or short period. In some patients the attack comes on suddenly. That is to say, they take to drinking large quantities of alcoholic fluids without any premonitory symptom. But more often there is noticed an alteration in character and temper that forewarns those who have anything to do with the patient. In the case of a married man, the wife can almost always tell

when an attack is coming on. The length of these attacks
vary greatly, more especially according to the duration
of the disease in the patient. In the early history of the
disease, the drinking bouts often last from one to three
weeks, and during that time the patient is constantly
drinking. When he cannot get the quantities of liquor
that he requiries outside, he takes to drinking in his own
rooms or house. Nothing will stop him. If his friends or
servants try to get him to leave off, he storms and rages,
and terrifies them into submission to his ways and wants.
His excuse for drinking is always that he is excessively
weak and nervous and requires support, and that it is absolutely necessary for his life that he should have stimulants.
His appetite soon disappears, and he only makes vain
efforts to partake of any food that is brought to him.
Great sleeplessness and restlessness comes on, and in fact,
the patient is often on the verge of delirium tremens when
the disease abates either gradually or suddenly, and he gets
fairly well. When it ends suddenly, it is from an attack of
acute or subacute gastritis, for which he requires and seeks
medical aid. The craving for drink having also disappeared, he willingly submits to medical direction, and under
judicious treatment, recovers. When the attacks go off
gradually there are less severe gastric symptoms, and the
craving having become less, there is a diminution in the
gastric and nervous troubles.

After patients have lived for several years with these
periodical attacks, the duration of attack diminishes in
length, and they increase in frequency ; the cause of this
being chiefly due to the effects on the gastric system. The
stomach much sooner resents the large quantities of alcohol
put into it, and consequently the drinking fits are cut short
by attacks of gastritis, and often also enteritis. But from
the attack being shorter, the interval of diminution in drinking also becomes shorter, so that the patient gradually goes
on from bad to worse. Once the drinking fit passes off, the

patient generally expresses great horror and grief at his propensity and the effects it produces, and will make all sorts of promises to abstain altogether from alcholic drinks. If the case is not a severe one, and the moral surroundings of the patient are good, he will often keep his resolution for a long period. But eventually, from some cause, either of social temptation or mental worry, he again breaks out and becomes wholly ungovernable. We have known cases where a man having recovered from a periodic attack would go for a period of from four to eight months without drinking any alcoholic liquor, but the first taste of liquor after that abstinence would bring on an attack.

A great many cases are called periodic which are really chronic, but with temporary exacerbations, and we will discuss these under the chronic variety. We only include under the periodic those cases where there are complete periods of a natural condition of mind and body ; and in almost all these the patients are either total abstainers or extremely careful and temperate in their habits in the intervals between the drinking fits. The higher the moral nature of the patient, the longer are those tranquil intervals, but the lower the moral nature the shorter are those periods. The reason being that a man of high moral nature, either from culture or inherent preception, gets more control over himself and will battle long against the craving when it is coming on, although eventually he gives way not from any fault of his own, but on account of his being the subject of a mental disease entirely beyond his control. On the other hand, a man of low moral standing will never try to keep himself from the temptations that he knows are dangerous for him, and never makes any attempt to stave off the craving, but yields at once. These last are a much more hopeless set of cases to deal with, and are almost always incurable, and eventually die from accidents or diseases induced by their habits.

Dipsomania that comes on from injury to the head is

generally of the periodic variety. Blows or falls on the
head are followed at a long or short period afterward by
periodic attacks of drinking mania. In surgical and medical
practice it is well known that similar blows on the head are
not infrequently followed by epileptic fits, and both these
classes of cases, viz., the accession of periodic attacks of
dipsomania after injuries, and the periodic attacks of which
we have been treating at length above, tend to point to a
relation between epilepsy and dipsomania. Just as an injury
to the head by producing some alteration in the nerve-
centers brings on epilepsy, so do they at other times bring
on dipsomania by no doubt causing also some alteration in
the nervous tissues. Again, where there is insanity in a
family, in the progeny there is a great liability to a repetition
of that same condition of the nerve-centers which produces
insanity, or else some modification of it, such as epilepsy or
dipsomania, for in almost every case of periodic dipsomania is there to be found a history of nervous affections.
There are cases on record of transitions between these two
affections. Either an attack of drinking passing off by an
epileptic fit, or patients at one time having an attack of
epilepsy and at another time an attack of craving for drink.
In fact, the two affections are very much alike in many
ways. Judging from cases under observation, the periodic
variety seems hardly ever to be induced by acquired habits
of drinking alcohol, it being almost always the result of
strong hereditary tendency to neurosis of some kind. Our
object in laying such stress upon the real insanity of dipsomania is to get its recognition as such by the legislature.

We stated just now the difference in effect upon the
course of dipsomania caused by the presence of more or
less high moral nature in the patient. All those who have
studied insanity have come to the conclusion that very
often there is a complete absence of the moral nature—
patients not having the slightest conception of any moral
obligations due to their fellow beings or themselves.

Very frequently, among dipsomanics, is there this complete absence of any morality.

It is not often recognized. Friends and acquaintances of the patient accuse him or her of immorality unjustly.
• For a man cannot be immoral who has no ideas of morality to start with. Many will think such a state of things incredible, but it is really not so. Patients suffering from dipsomania often behave very well for a time, but it is only from the effects of habit from the way they have been brought up. They may do what is right, and often appear to go out of their way to do what is right, but if one could get at the root of their actions, it would be found that they were not guided by any moral feeling or sense whatever.

Dipsomaniacs show this perversion, or rather, absence of morals, in almost all their dealings with their friends and relations, and it is on this account that the disease is such a scourge to the friends and relatives. The patient will give the most solemn promises not to take any stimulant, but the backs of those to whom he has made the promise are no sooner turned than he violates it, either by obtaining the liquor himself or not refusing the first or any temptation that is set before him. They always have a plausible excuse when taxed with their violation of promise. They will also prevaricate most cunningly until the question is put directly to them. It is curious that in spite of the absence of moral nature they will rarely tell a deliberate lie. Some people might think this was due to a vestige of morality, but it is, we think, due simply to a knowledge of their weakness, and to the power exercised by a stronger mind over theirs. This habit of prevarication, of never giving the lie direct, they are cunning enough to bring forward as a plea for their being trusted occasionally. But in our opinion, a dipsomaniac can never be trusted. It is not right that he should be led to suppose so, for that would do him inevitable harm, as we shall see presently in discussing moral treatment for dipsomania, but at the same time

one would never rely on the promises given by a dipsomaniac, more especially if it is in regard to taking any precautions for guarding against his propensity.

As regards the recurrence of these periodic attacks,
they do not seem to have any regular intervals. In some
patients there is a cessation of the drinking propensity for
many months, in others, only for a few weeks. In the
chronic variety we shall have to mention that where periodic
exacerbations occur, they often do so at regular intervals,
and the length of time is generally a month.

In women, where generative functions are periodic, dipsomania, like other varieties of insanity in women, is apt to
recur at intervals corresponding to the menstrual periods.

The progress, then, of periodic dipsomania is generally
an increasing frequency of the recurrence of the attacks
which may verge into the chronic variety, but as often as
not kills the patient before that stage is reached.

Among this class, however, are found some who do
appear to be cured by careful and judicious treatment
during the intervals. The treatment consisting in elevating the moral nature by every possible means, improving the patient's general health, and getting him to employ
his mind and body actively. There is one drug which
ought to have extensive trial for treating the periodic
attacks. Bromide of potassium is well known to be of
great service in reducing the frequency and force of epileptic attacks, and seeing the close relationship of dipsomania to epilepsy, it ought to be of use. We have not
had the chance of giving it much trial in the periodic
varieties, but in the chronic it is of immense service.

In those cases where the moral nature is evidently low,
weak, or absent, no hope can be held out for cure, and
the patients will be an endless source of misery to themselves and relatives till they die. It is for these cases that
a proper law is so much required. During the attacks
nothing but restraint can keep them from liquor.

Men who suffer from periodic attacks live a little longer than those whose attacks are chronic in their nature. Among women the periodic is more common during the whole of the time of the existence of the generative functions. Chronic dipsomania in women more often comes on later in life.

Chronic Dipsomania.—We are certain that this is the most common form. Those suffering from it generally commence drinking at an early age, usually 17 or 18, and by 30 years of age most of them have died. They die either from the direct effects of the drink they are continually consuming, or else from diseases induced by it, such as absolute insanity or diseases of liver and kidneys. The disease is not generally recognized in the first few years of its course, as the patient's friends and relatives merely suppose him to be living a little too fast, and that he will pull up soon. However, after a few years it is found that he is never really free from the effects of liquor. He is hardly ever really drunk, but is never sober. As we have said before, champagne is the usual drink during the first years of a dipsomaniac's career. When he takes more than enough, it shows its effects by making him very sleepy, and he will lie down at the most unusual times and places, and fall off into an apparent dead sleep, from which it is very difficult to awake him, and if one does so, he usually manifests a disgusting temper, and will curse and swear at his best friends. This sleep does not rest the brain at all, as when they do wake up voluntarily they are by no means refreshed, and immediately have recourse to another stimulant, which generally freshens them up for a short time, but they very soon require re-stimulating.

In the early days of their career, these men often go through a large amount of fatigue, and consequently for a time ward off the mental effects of their drinking. They will hunt vigorously and shoot, walking long distances, but the excitement of horse exercise is what they most delight in, as the fatigue enables them to drink large quantities of stimulants without much apparent effect. But after several years the muscular as well as mental system begins to fail. They cannot go through the fatigue they were formerly capable of. Their will also fails. Overnight they will make engagements and promises for the morrow, but the morrow finds them unable to perform their promises, nor have they the will to carry their project out.

Every now and then the patient acknowledges that he is drinking himself to death, and he will slacken in and reduce his quantities of stimulants, but never leave off entirely. He very soon, however, resumes his usual large potations. Occasionally he can be induced to go to a home or inebriate asylum for a time, but he never stays very long, and is no sooner out than he resumes his former habits. It is a fearful existence, the constant craving for fluid, and that fluid must have some taste. All dipsomaniacs have a complete antipathy to water. They will drink anything but water. If they cannot get stimulants they will drink soda-water, lemonade, or ginger-beer, in enormous quantities, so much so indeed that the mere quantity of these apparently harmless fluids which they drink is quite sufficient in itself to ruin their digestive powers. Many people think that it is only necessary to substitute these supposed harmless fluids for alcoholic drinks to produce a cure. The idea is entirely fallacious, as large quantities of those fluids are as likely as not to induce dipsomania itself. There are two classes of dipsomaniacs that are very different from each other in their habits of eating. One set has the most voracious appetite, eating enormous quantities of meat, cooked in all forms. These patients are generally wine-drinkers, and very often claret-drinkers. Another set of patients eat very little, their stomachs being unable to retain or digest the food. In the first class the large quantities of food taken is not properly digested, as is evidenced by their constant call to get rid of their excrement. In fact some of these patients seem to occupy a great part of their time by evacuating their bladders and rectums of the large quantities of food and drink they ingest. The diet of those who only eat a small quantity is generally a meat one.

Their usual routine of life consists in getting up early or late in the afternoon, rarely in the forenoon. But during the morning they have been taking continual small doses of beer, champagne, hock, or claret, generally diluted with soda or seltzer, and have short snatches of a boozy sleep. Having come out by the afternoon they attempt a dinner in the evening, and with a fair amount of liquor in them, they bring themselves up to the scratch to play billiards or cards, but they soon get tired, as they call it, and lie down and go off to sleep. By about one or two in the morning they wake up, and from then till eight or nine is their miserable time. They cannot sleep. The effect of the liquor is passing off and they get frightfully nervous. They often wander about the house, and if they cannot obtain any liquor before the general hour of rising of others, they are usually reduced to a state verging on delirium tremens. Directly they get some liquor, however, they go and undress and go to bed, and so on till the afternoon, when the repetition of the night before takes place.

In the early history of dipsomaniacs they have usually had one or more attacks of delirium tremens, but after recovering in a way from them they are never able to drink the large quantities of stimulants they used to do formerly, as they get so much sooner prostrated and intoxicated that they never get real delirium tremens again. We have seen patients in whom the disease is of long standing so altered that one or two glasses of beer would make them quite boozy. Also, after the attacks of delirium tremens, they never seem to recover their muscular tone and power, and consequently cannot go through the same amount of exercise. The reason of this is of course that they go on drinking.

If their liquor was stopped, physically they would get quite well. A characteristic sign of the chronic dipsomaniac is a diffuse roseate hue of face and neck and a watery aspect of eye. They are generally full-bodied, but they are always weak on their legs, and late in the disease get into almost a state like locomotor ataxy. Besides the roseate hue of face there is another symptom indicating the determination of blood to the head, and that is a constant recurrence of slight hemorrhages from the nose. The pocket-handkerchiefs of dipsomaniacs always contain secretion mixed with blood. This is only found in old standing cases, and is often alone due to the congestion of the liver, which is constantly present. The urine in the morning is generally very high colored. The pulse in the morning is also always weak, soft, and small. We have already stated our belief in the absence of any moral feeling in these patients. It is hard often to decide whether the drink may not have removed their moral sense by its effects on the brain. For a man may have possibly started in life with a fair proportion of the moral sense, but the physical action of alcohol on his cerebrum may have destroyed his moral appreciations. No doubt this occurs in truly acquired dipsomania. But in these latter, the disease comes on later in life. Where the disease is hereditary, and is an insanity, and begins earlier in life, one generally finds that even in their boyhood they evidently were devoid of moral sensibility. The cunning of these patients in using every artifice to get drink, when they know efforts are being made to keep them from it, is really wonderful. They will hide bottles of spirits among their clothes. They will, in the earlier stage, walk long distances so as not to be seen drinking, will use persuasions and threats to those attending on them, and will eventually beg anybody for drink to satisfy their craving.

The symptomology of inebriety is far more familiar to the reader than its philosophy and causation, hence it is omitted for want of room.

CHAPTER V.

Men in the world can be very readily graded, as they
ascend from the lowest to the highest types, by two very
essential qualifications—the one, their degree of *intellectual
sense ;* the other, their degree of *moral sense.*

By the lowest type of men we mean one both ignorant
and immoral ; by the highest type, we mean one both intelligent and moral These two qualifications can, to a considerable extent, be separated : we often speak of the *mental*
and *moral* qualities as distinct ; and we recognize the fact
that in the same individual, frequently, the levels of these
two qualities do not correspond. For instance, we can have
one person, who will grade *higher* in his morality than in his
intelligence, and another, *lower* in his morality than his intelligence. The rule, however, is,—the ability to think is
accompanied by an equal ethical ability.

In the gradual advance of a race from savagery to civilization, their progress is occasioned by and marked by gradual improvement in both these particulars.

In any man, his intelligence and his ethical ability are
raised to a level corresponding with his previous practice in
performing these kinds of action. The savage, compared
with the more advanced man, is inherently less *able* to perform these kinds of thought, and as he progresses upwards
towards civilization, through generations, there is a gradual
improvement in his *ability,* until as a most advanced man he
finally has most *capacity* to *think* and to *do right.*

39

The first essential in an advancing man, or advancing race, is *activity*. An indolent, idle man or race never advances. It is the activity or exercise of the brain that increases its ability. Accomplishments and excellences are acquired only by practice and exercise.

The rapid competitions of active society, therefore, necessitate *activity* on the part of the individual to avoid suppression and elimination. Hence, improvement of individuals is most rapid in active society. The continuous brainwork, which under such circumstances becomes a necessity, improves the thinking ability of the rising man.

In the rising man, because it is a necessity, his intellectual (competive) ability antedates to some extent his ethical sense. In a society which is advancing from a savage to a civilized grade, competitive ability is acquired *before* general communal interests require the harmonizing of the competitions. The safety and the welfare of society soon demand that the competitions of its membership shall be harmonized. This gives rise gradually to the evolution of higher and higher rules of conduct. Public opinion and moral sense, with laws and government to enforce them, thus become of a higher and higher grade, until the highest type is reached.

In this gradual improvement of society, the advancing man is more and more practiced in ethical observances, until, as an accompaniment to his improving ability to compete, there also grows in him a better ethical sense of the rights of others. A high ethical sense may well be called the *capstone* of human improvement.

In the best communities to-day we find some individuals —in most communities they are few, in no community do they reach a majority—who have a *high* order of ethical sense *inherent* in them. These persons are always inherently intelligent, for it takes the highest order of intelligence to so understand the complexities of civilized competitions as to be able to formulate and maintain high rules of conduct to suit them. Such a high order of man—highly intelligent and highly ethical—is only the result of a long line of generations of this kind of practice.

We spake of civilized countries, and of civilized societies as though their membership were uniform. This is not so. Every community, even that graded highest, will furnish examples ranging all the way down from the highest type, just described, to the lowest. In any community, in a so-called civilized country, we can find persons who inherently grade very low in thinking ability and in ethical sense. They are at the savage level. A great many have the savage level of moral sense, while they hold a higher level in intellectual sense. This, in my opinion, is particularly the case with deteriorating or degenerating individuals, they have lost their ethical sense in advance of their intellectual, —the latest evolved and most delicate goes off first.

We have not yet invented a cerebral *dynamometer*, by which we can test and record a man's intellectual capacity or his moral strength, but in our associations with others we instinctively recognize such information to be very valuable. It is interesting to note how much we are engaged in this very kind of work. We are continually making estimates of this character, and it is astonishing, in a crude way, how expert we become at it.

In making our estimates in society, the position a person holds is one to which he has *arisen* from a *lower* level or one to which he has lapsed from a higher one. Human brain ability is not by any means a constant quality—there is no standstill level. The index rises and falls in the course of the life of the person, and also it varies in the course of the line of descent. In so-called civilized society the *lapses* probably constitute the large majority of the incompetent.

Ability, at whatever height it occurs in an advancing man, is raised to that level solely by previous brain practice. The practice is either personal or ancestral. The person

receives his ability at a certain level from his ancestry and
raises it or lowers it by his habits of thought. By far the
best and most stable ability is that which is the result of
practice reaching through a long ancestral line.

While it is true there is only *one* way of improving brain
ability, namely, by activity, there are *several* ways of lowering it. There are several ways in which the *lapsed* members
of society have reached their levels.

Brain inactivity, I have already stated, is the *physiological*
method of lapsing. There are a number of *pathological*
methods.

Our pathology is always injured or impaired physiology,
so I have dwelt this long upon the physiology of cerebration,
in order more properly to approach its pathology.

Whatever injures the structure of the brain impairs its
functional action. This impairment is exhibited in altered
conduct, in a loss of ability, both intellectual and ethical.

We can go through the wards of our insane asylums
and find numbers of men, who once ranked high in intelligence and in morals, now lowered in both. Defect, injury,
or disease now renders the brain of each of them *less able*
to perform at as high a level as it formerly did.

Insanity, indeed, is only a name we give to a certain
degree of brain incapacity. As generally defined, it simply
means there is such a degree of incapacity as renders the
person an *unfit* member of society. For this reason, for his
own or his neighbor's safety, he has to be placed under for-
cible restraint; his brain is so lowered in intellectual ability
that he is unable to compete for his living, so a support has
to be given him; and it is so lowered in ethical ability that
he is a nuisance or a danger to others, so he has to be
restrained.

In society, short of the degree called insane, we have
innumerable varieties and kinds of disability, exhibited by
the peculiar, the cranky, and the delirious.

Besides the long list of diseases, traumatisms, and

defects of the brain, which impair its functional capacity, we also have a long list of drugs, which, taken into the circulation, bring down the brain's capacity to a lower level.

I need not go over the long list, nor point out the peculiarities of their actions, but at once mention *alcohol* as one of them. This agent, from its general use, brings down more brain capacity in the world—and it certainly does a great deal to produce the *lapsed members* of civilized society.

If it were not for the fact that alcohol has in the system a special effect upon nerve centers, particularly high brain structure, men would never have used it as a beverage. In seeking and taking it the alcohol drinker is after its *brain* effect.

The brain is the organ of thought and all conscious action. The partially hardened condition of its delicate structure that alcohol produces, renders it less capable of cellular motion or action. Its conscious sensitiveness is lessened thereby. It is less able to feel. The alcoholized man therefore says he " feels better " or " feels good," and acknowledges that this kind of lowered sensitiveness is what he likes. He " feels better " if his brain feels less any discomfort or pain he may have. Even in health, if he have no special discomfort, the benumbed condition is a more comfortable one. The well man therefore says he " feels good." The *luxury* of alcohol drinking consists in this condition of the brain.

I wish to draw attention to the fact, that in order to obtain the pleasant, comfortable condition of lowered sensitiveness, the alcohol drinker does not avoid or fail to have the alcohol effects on his brain—there is a general lowering of function. When the brain's ability to feel is lowered, its ability to think and to adjust conduct ethically are also lowered. The keenness of a high ethical sense is probably the *first* thing blunted.

The disability of the alcohol drinker will vary according to the *amount taken ;* according to the *inherent strength* of brain structure ; and according to the *length of time* the brain-abuse is continued.

For instance, first, according to the *amount taken*, conscious sensitiveness lets down from a slightly benumbed, comfortable condition, to that of complete anæsthesia ; intellectual ability varies from being " a little off " to the condition of a temporary dement ; the ethical sense varies from slight indecorum to full viciousness or madness. Secondly, the degree of brain disability under alcohol will vary according to the *inherent brain strength* of the drinker. Weak brains will be lowered in ability more than strong ones, and low grade savage brains, or defective ones, will exhibit their incapacities in the lines of their deficiencies. Thirdly, the disability will also vary in proportion to the *length of time* the drain-drug abuse is continued. A single debauch can be fully recovered from, but long-continued use produces such an injury the full function is never restored.

Sometimes all three of these conditions obtain in the same person ; when this is the case, the degeneracy of brain function is extreme. The *inebriates* constitute this extreme class, always degenerated to a very low level in intellectual sense and ethical sense. Of course there is a *tendency* to return to previous capacity when the brain-drug abuse is left off ; but I question whether an inebriate's brain is ever fully restored to its previous capacity, or ever reaches the level it would have occupied without the injury.

As society advances from a savage to a civilized level success depends more and more on brain strength. The most advanced society pays the highest premium for brain ability. In active civilized society, the safety and success of the individual depend almost altogether upon this qualification. In savage society, the muscle strength suppresses the weak, in advanced society it is the brain strength of the more competent that suppresses the incompetent.

The drinking man lets down in business and loses his

money because his thinking capacity is lowered by his habit —his ability is weakened. The drinking man also lets down in morals, *falls* into vicious habits, because his ethical sense is weakened.

In modern society, brain idleness probably puts most men into the *eliminating* level ; next to idleness comes alcohol. These two agencies rapidly rid crowded society of their unfit membership. Nothing is so rapid an *eliminator* as alcohol. The least fit, both in the idle wealthy ranks, and in the idle poor ranks, are most given to its use. Under the light of advancing science, the use of alcohol is becoming more and more confined to the class of the weak-brained and the vicious. The intelligent, for the sake of maintaining this intelligence and superior fitness, are learning to leave it off.

Society is much interested in the intellectual level of its members, but it is most interested in their *ethical* level. The safety and survival of the individual are most dependent upon his intellectual ability ; society, though, is principally interested in his ethical ability. The safety and welfare of society depend upon the moral status of its people. The good of society demands this.

Communities, races and nations, like individuals, are engaged in competitive life. The most successful is the one which has the high intellectual, progressive capacities of its people welded into a harmonious, united whole by a high ethical sense.

There is a very satisfactory scientific explanation therefore to the fact, that the most altruistic persons, those most interested in public good, have always been opposed to alcohol drinking. They recognized the fact that it lowered the intellectual and moral abilities of the people and tended to weaken and disintegrate society. Probably one reason they have never succeeded better in enforcing their opinions upon the attention of the alcohol users, is because they have not had the advantage of recent scientific knowledge to back their instructions—the brain has been left out of their philosophy altogether.

I have often been make impatient in listening to the lecturer presenting the " scientific aspects of the alcohol question" to an audience, to see him illustrate extensively with charts and spend hours to show the effects of alcohol upon the coats of the stomach, and upon the structure of the liver, the lungs, and the kidneys, and never allude once to *the brain*, when the fact is, alcohol's principal effect is upon this organ ; and the functions of this organ so far transcend the functions of all the others, that, I might say, there is no comparison.

When the individual in society is taught the fact, which he seldom knows, that alcohol incapacitates this very organ upon which his safety and success in the competitive world depend, he will be very much less inclined to use it. And when society recognizes that with alcohol a low-grade, vicious man can be made chemically out of its excellent man, and that this process is continually going on among all its ranks, it will be more alive to spread scientific instruction upon the subject.

CHAPTER VI.

INEBRIATE DIATHESIS AND ITS CONDITIONS AND RELATIONS.

By the *inebriate diathesis* is meant that constitutional proclivity, or *neurosis*, which impels to the inordinate use of
narcotics.

This includes the hurtful consumption of opium, chloral,
cocaine, etc., as well as of alcohol. The latter, however,
will be more particularly the subject of the present inquiry.

The peculiar *bent* of the constitution herein referred to
has been classed as a specific mania ; and has been called
by Dr. Norman Kerr and others *narcomania.*

In the interest of brevity, I will begin what I wish to
advance on this subject with a homely illustration. We are
assured by anatomists that the bones of the skeleton—it
matters not what shapes they assume—are, as to form,
modifications of a single fixed type. This type is represented by the separate and distinct bones of the spinal
column.

When we attentively consider the group of nervous
maladies known as the *neuroses*, we perceive that they bear
a close relationship with one another. We learn that
either member of this family may assume the characteristics and features of some other member. We become
aware of the fact that, seemingly, one of them may be
transformed into the semblance of some kindred neurosis.
We see that this transmutation may be effected through the
operation of heredity—a neurotic form in ancestry appearing as a different form in posterity. We are also impressed

that this interchange may even take place in the same person ; as when epilepsy, or dipsomania, gives way to amnesia, or some other neurosis.

While it is reasonable and necessary to study the bones of the skeleton separately and individually, in order to arrive at right conclusions respecting the body as a whole, it is also philosophical, and, indeed, essential to a correct apprehension of the subject, to study with care the distinct *type* of which these bones may be modifications. It is desirable to examine the spinal bones, so that a proper estimate may be made of the nature of the modifications assumed by various parts of the skeleton—and the reasons of them.

The true and perfect *type* of which the entire assemblage of neurotic maladies is representative, is, it appears to me, simply *epilepsy*. The varying aspects which this disease assumes, its several grades of intensity, as well as its origin, —so far as that is known—seem to point it out as the great central source of the several neurotic besetments. If this be true, even in a general sense, it follows that in seeking the origin of the inebriate diathesis an examination of the causes and phenomena of epilepsy will be a necessary work.

Epilepsy has been classified as either *centric* or *excentric*. That is to say, it arises sometimes from causes situated within the brain ; and at other times, from causes exterior to the brain, but influencing the condition of that organ. These causes, while often attended by pain, may operate without exciting actual pain. In epilepsy there is a constant, unrelenting *irritation*, nagging the great nervous center. This irritation may be simply a morbid impression acting through organic sensibility, or it may be attended by actual pain and distress.

Hence teething, indigestion, and worms excite convulsions in children, which are epileptoid. So, too, tapeworms, affections of the liver, or stomach, or kidneys, and the like, may excite true epilepsy in the adult. It is obvious that epilepsy from such causes is more or less amenable to treatment. But the treatment of *centric* epilepsy—frequently hereditary, with its proximate cause within the brain—is more difficult ; and usually the most that can be expected is a certain toning up of the general constitution, by which means the effects of the original evil may be better endured. Through such measures some hope may be entertained, perhaps, that the violence of the epileptic seizure may be abated. In short, some slight possibility may appear, that, instead of true and complete epilepsy, there may be substituted amnesia, or neuralgia, or some other and milder one of the neurotic forms.

An educated and refined gentleman—now a retired clergyman—once suffered with repeated and severe attacks of epilepsy. The coma continued many hours. The epileptic form disappeared under treatment ; but it was followed by seasons of absent-mindedness (amnesia), lasting for days. No recollection of the life of this neurotic trance was or is retained. The same individual suffered afterwards for nearly two years with excruciating pains. They were referred at first to the stomach, cancer being feared. Subsequently they were located in the pleura. Finally the trouble was attributed to biliary calculi. The "cure" was complete, sudden, and unexpected. No calculi ever put in an appearance. The physician in attendance upon the case in its neuralgic form had no accurate knowledge of the previous history of his patient.

Central or *centric* epilepsy is often hereditary. But that fact does not militate against the idea that *irritation* is the *primum mobile* of the disease. The misshapen head, the undeveloped brain, may readily be supposed to act as sources of unceasing irritation. Injuries to the head producing epilepsy are too common to call for remark. Knowing, then, that irritation is an efficient cause of epilepsy in all cases where its cause has been satisfactorily determined, we are justified in assuming that in those cases where the source of the malady does not plainly appear, the high probabilities are that some undiscovered or masked point of irritation is nevertheless operative as the proximate cause of the disease.

I say *proximate cause*—for it is evident from the fact that irritation does not always produce epilepsy, that, after all, there must be some peculiar predisposing constitutional susceptibility in epileptics that is not found iñ people generally. And here is the key which unlocks at least some of the secrets of heredity. It is the peculiar impressibility of the epileptic constitution, which, being hereditary, makes possible the heredity of the disease itself—circumstances favoring its full development being presented. It also makes possible the heredity of those neurotic forms recognized as kindred with epilepsy. This constitutional susceptibility therefore explains the heredity of inebriety, hysteria, chorea—epilepsy itself. It explains the heredity of dipsomania, and all the features of the inebriate diathesis.

When there is present a peculiar constitutional tendency allied to the epileptic diathesis, then any irritation of nerve—if severe and unremitting—will be likely to develop some form of neurotic disease ; and this will correspond, not with the nature of the irritation itself, but with the nature of the constitutional trend. Dr. Cheyne refers to a case of epilepsy "which was caused by a cartilaginous tumor of the size of a large pea, which was situated on a nerve. Upon excision of the tumor the fits ceased." Why may not inebriety—narcomania—be developed by causes as small and apparently as trivial ?

We are admonished to look for the pathological condition provocative of inebriety. Why not insist upon some invariable pathological condition in explanation of epilepsy ? Very often, indeed, this morbific cause cannot be determined ; but when it does appear, it is always found to consist of some point, or points, of unflagging and remorseless *irritation.* It may be repletion, or it may be famine and weakness. It may be too much blood, or too little blood. It may be observed in the center of the circulation—the heart—or it may reveal itself in irregularities and disturbances of the equilibrium among the arteries, veins, and capillaries. But whatever else it is, or wherever located, it is *irritation* worrying and exhausting the nervous powers.

Now certain constitutions bear up through the changes of life fairly well, until some serious injury overtakes the physical organization, such, for example, as a blow upon the head, a wound in battle, or even a long and trying illness. Irritation at once begins to do its work. The ordinary and natural constitution gives way. It is weak, exhausted, weary. It has become unequal to the requirements even of ordinary life. It reaches out for aid, or rather for rest and repose. The inebriate diathesis is established, and the anæsthetic—the lethal influence of narcotics, and especially of alcohol—is invoked. The call is not feeble and uncertain, but earnest and reckless.

An habitual drunkard "shot a bar-keeper and set fire to a saloon, without the occurrence of any quarrel or dispute. A homicide was hung. A *post-mortem* revealed a splinter of bone that had pressed upon the brain for ten years, the result of a blow upon the head." The drinking habit began shortly after the infliction of the injury.

A daily paper of Cleveland, Ohio, dated October 13, 1886, contains a pitiless notice of the downfall of the Rev. ——, a clergyman living in the city. " In the gutter again," was part of the heading. The account says the reverend gentleman was sent to the workhouse, and also contains a notice of the dismissal of the deliquent from his position in the church, with a warning to the Christian public to beware of him. On the 6th of October, 1891, after the lapse of about five years, the same daily journal contains a sympathetic account of a *post-mortem* held upon the body of the

reverend gentleman. This was had in an insane hospital in
which the unhappy man had been placed. This *post-mortem*
disclosed the fact that the man was carrying in his lungs a
bullet weighing over one ounce, received while bravely
fighting the battles of his country in the war of the rebellion.

It is evident that the treatment proper for inebriety must
occupy a very wide field, as its proximate causes are so
varied. It is filled with innumerable facts, requiring the
utmost skill, both in observation and discrimination. In its
very simplest view, three considerations must be kept before
the mind :

    1st. The causes of inebriety.

    2d. The nature of the drunken state.

    3d. The functional disturbances and physical degenera-
tions that are sure to follow long-continued habits of
intemperance.

The Diathesis and Cachexia of this affection has never
been carefully studied, but, standing on the border, we
catch intimations of its vast possibilities, awaiting the
future investigator.

A diathesis implies a special predisposition to certain
kinds of diseased action. A cachexia is a developed diathesis, a real condition of disease of the blood and tissue.
A diathesis is a more or less remote cause, but a cachexia
is a present part of disease, an influence on the organism.
Using these terms together, certain well defined conditions
appear to be traced, from which inebriety can be studied
and prognosed with much certainty. The first and most
common form seen may be classified as the nervous diathesis and cachexia. They may be noted by their spare
bodies, light hair, changeable countenances and restless
animation. Easily agitated, intense functional activity, so
sensitive as to show signs of mental disorder when excited.
Mentally their views of life are unreal, visionary and
exceedingly changeable ; seldom calm and deliberate, but
abrupt, jerking, and when exhausted, prostrate. When

this is inherited it is often accompanied by a vain, weak mind, with boundless self-esteem, always on the verge of eccentricity.

The mind dashes from one extreme of emotion to another, either showing excitement that is morbid, or degrees of feebleness that are abject. Women are often of this class, and persons from the wealthier circles of social life. Luxurious habits, indulgence of whims and fancies in childhood, are active causes of such a diathesis. This is a distinct type of mental defect from which inebriety and insanity are an almost constant sequence. An early diagnosis and prognosis, with treatment, would furnish the only practical means of escape from this diathesis. Another condition may be noted, called Strumous Diathesis and Cachexia. Here the patient has diminished vital energy, imperfect nutrition, with mental weakness and perversions, unnatural special developments, which have a constant tendency to exhaust themselves, as, for instance, in case of extraordinary mental or physical genius, where the body is out of all proportion. Many of the so-called great men of the world belong to this class. They are nearly always anæmic, complain of frequent headaches and have often a tuberculous history, accompanied by a species of degeneration which has a constant tendency to concentrate in some particular organ. Any exposure to cold takes on an aggravated type of symptoms, or injury to any organ assumes the most alarming phase.

The nerve centres are always more or less involved, and shocks or injuries are acutely felt. Inebriety appears as a part of the natural history of this diathesis. Paroxysmal drunkards belong to this class, and of all others they are the most thoroughly demented during the attack. In many cases an ARTHRITIC DIATHESIS and *Cachexia* are well marked. Here a chronic state of imperfect nutrition, with mental exaltation, or depression existed. In many cases a history of rheumatism and deposits about the heart, with calculi of gall duct and bladder were given. The circulation was weak, and the intellect unsteady. A capricious, irregular appetite accompanying these conditions often forecasts inebriety, and generally precedes violent attacks. A

SYPHILITIC DIATHESIS AND CACHEXIA

are known by temporary headaches, amaurosis and flying pains, with mental and moral defects of various grades ; anæmia and affections of the heart are often present. This, with a clear syphilitic history, are evidences beyond question. In this, diathesis inebriety or insanity develops from the least exciting causes. There is a *Hemorrhagic Diathesis* and *Cachexia*, which is characterized by anæmia, blood deficient in fibrin, suffering from persistent neuralgias, and a constant tendency to copious hemorrhage from the slightest causes. Inebriety often follows from this condition. Two cases have fallen under my notice where this diathesis ended in uncontrollable drunkenness and death. A *Fibrinous Diathesis* and *cachexia* may be the active cause of inebriety by the constant tendency of the blood to exudations of unhealthy plasma, and general hypertrophy and induration of the cerebral system, all the functions being languidly performed. In all these cases the normal equilibrium of the nutritive functions is destroyed, and inebriety follows as the first stage of degeneration. These are some of the many conditions of the body coming down from inheritance, well marked, and so often preceding this affection, as to be called active and predisposing causes. These conditions are not always distinct, but blend into each other, multiplying and diverging.

CHAPTER VII.

GENERAL PREDISPOSING CAUSES.

A study and grouping of the general causes will afford data, from which we can understand inebriety in its first manifestations. The first group, called General Causes, are those which frequently produce inebriety in common with other insanities. Of these *mechanical injuries* seem prominent, such as injuries or concussion of the brain and spinal cord, and consequent alterations of nutrition. Blows on the head are not unfrequently followed by violent paroxysmal drunkenness; railroad accidents, where the concussion and surprise are sudden and overwhelming, causing intense re-action on the nervous system (producing at first insignificant physical lesions), often develop inebriety and mania. A chaplain in the late war, injured in the leg by a shell, although but a simple flesh wound, was several months in the hospital before recovery. From this time he became an incontrollable drunkard, and eventually died. A prominent physician was precipitated from his carriage on the head, and received flesh wounds of a minor character. Very soon after inebriety was developed, and he ended his life a raving maniac in an asylum. The rule in many cases indicates that injuries of the brain, spinal cord, or any part involving numerous nerve trunks may end in cerebral disturbance, of which drunkenness is a very common stage of the lesion, the injury being small in comparison to the mental disturbance which follows. Peripheral irritation or reflex excitability has been noticed as a common cause in many cases. In a letter to the

55

Psychological Journal, a physician describes two cases where tape-worms excited inebriety and mental hallucinations, which disappeared on their expulsion. Several instances are on record where the use of pessaries in prolapsus uteri and the prolapsus alone, has been the cause of mental disturbances of which drunkenness was the most prominent symptom. The irritation from prolonged lactation, or in dysmenorrhœa, ammenorrhea, nymphomania and functional disturbances of the genital organs, are frequently the beginnings, and in many cases the active causes of this disease.

A case of chronic masturbation, which ran into inebriety, came under my notice lately. Numerous cases of puerperal insanity, preceded or followed by inebriety, are recorded in the English obstetrical journals. Cases of drunkenness during pregnancy, ending at delivery, are not uncommon. A remarkable case was noted at the Dublin Lying-in Hospital, some years ago, of a woman, who during the period of lactation, drank gin to excess. This craving for stimulus began when the child was born and continued until the death or weaning of the child, the former being the most common termination. When lactation ceased the desire died away. Another case of equal interest is reported in a western journal. A soldier, wounded at Pittsburgh Landing, previously temperate, became a paroxysmal drunkard. A portion of the shaft of the femor was injured, and continued to exfoliate dead bone until 1873, when it was removed by an operation. From the healing of the wound his drunkenness disappeared, and he is now one of the most enthusiastic of temperance men.

Previous diseases are also active causes. Dipsomania, erotomania or monomania often are associated with epilepsy in some of its obscure forms, as natural stages in its progress. Epilepsy, with its disordered fancies, auras and impulses, ending in apparent recovery, only to be followed after time by a repetition, closely simulates the impulsive drunkard.

The connection between inebriety and epilepsy is far more intimate than we are aware. Affections of the heart, either organic or nervous, and low chronic hepatitis, predisposing to hypochondria and melancholy, very often precede this affection. Neuralgia often ends in this way, by disturbing the mind centres and its equilibrium. Gout suppressed has brought on inebriety, but when it returned, drunkenness declined. Intermittent and malarious fevers have been followed by this disorder, and the medical attendant has been blamed for causing it, by prescribing stimulants. Persons suffering from acute rheumatism, after the severe symptoms have subsided, have exhibited the same affection. Diseases of the skin and renal disturbances have been noted as preceding drunkenness.

Diseases of an intermittent character (a hint of nerve complication) may be the primary cause of drunkenness, through the perversion of functions, beginning so obscurely as not to be noticed.

Dietetic diseases seem more frequently associated as causes owing to the intimate association between the nutritive functions and the nervous system. Dyspepsia, with all its train of perverted sensations and tastes, have often common sequence in this disorder. Excessive hunger and thirst are followed conditions which are often active causes. Indigestion, inflammation of the stomach and liver often leave a train of predisposing causes. Practically, any disease influencing or breaking up the nutritive functions of the body, has its first effect in perversions and cerebral disturbances, of which inebriety is a frequent result. Individual histories of neglect of hygienic care, followed by indigestion, dyspepsia, hypochondria, drunkenness and death, can be traced in every community.

Exhaustive intellectual and physical exertion, by breaking up healthy cerebral action, may develop insane longings. Over stimulation of the brain ends in structural changes and perverted tastes for alcohol or narcotics of some kind.

Scholars, and those who use the mind to excess, and laborers, who only exert the physical system, when exhausted from over work, have unnatural longings for something to restore the lost balance of mind and body. A prominent senator spent thirty hours continuously in the preparation and delivery of a speech, and became an inebriate from that time. The proprietor of an ice-house offered some men large wages to continue the work of filling the house ; being expert workmen they continued without much rest for fifty hours. Two out of five became drunkards, and dated it from this time.

Instances of similar character are not uncommon. Sudden excitement and exhaustion acts in the same way, one physiological effect is rapid contraction of the arteries, apparent in the pallor and consequent anæmia of the brain. The shock imparted to the senses through the mind, extends to the brain cells, producing similar commotion, often breaking up their normal action forever. The sudden loss of property, disappointment, trouble, unrequited affection, may so depress the vital powers and disturb the circulation, that drunkenness or insanity, or both will follow. Ambition to lead, where it becomes a dominant passion, filling up every moment of thought, lowers the nutritive functions and leads directly to perversions of taste. Religious activity, where the mind overflows with sentiment and emotion, may react into violent gluttonness or inebriety. No class of men are so thoroughly dyspeptic as clergymen and lawyers ; simply the reaction of the mind to restore the equilibrium broken up by irregular exhaustive labor. In every community large numbers of this class are on the verge of inebriety. The capricious appetites, and the excessive thirst for tea, coffee, cider, soda, lemonade, etc., etc., and the bitters and patent medicines consumed to cure ills that are imaginary, are all early stages of this affection ; a change of surroundings, and they become inebriates in the fullest sense.

Maudsley thinks that the depraved longings of inebriates are greater in the morning, particularly if following exhaustion and anæmia, owing to the contracted blood vessels during sleep, which have not had time to recover. Confirming this, a case under my care has strong cravings for liquor in the morning, which, if broken up, does not return until the next day. This craving for unnutritious fluids, or substances, generally indicates impending brain disorder. Excentricities of character are frequently associated with inebriety. The poets, artists, and scholars, and warriors of the world, who were inebriates, are examples. Their minds, dashing beyond the bounds of normal, mental and physical life, reacts in nutritive excesses. Of this class we might mention Alexander, Philip and Nero, among the warriors ; Shakespeare, Byron, Burns and Poe, of the poets ; and Johnson, Addison, Pitt, Sheridan, etc., etc., among the scholars.

An unbalanced mind resulting from a misdirected education is the common cause of inebriety. Where the laws of growth and elementary power are untaught, and the child grows up without any purpose or object in life, the faculties undisciplined, the appetites and tastes indulged, no system, only the gratification of the physical wants, no particular knowledge, the love of excitement constantly stimulated, and self-esteem encouraged, add to all this, predominant passions, extravagant and capricious desires, and drunkenness is almost sure to follow. The present system of cramming and over-stimulating children, in the study of imperfect text books, in worse surroundings, lays the foundation for both physical and mental dyspepsia, hypochondria and dipsomania. Two-thirds of our graduates leave the schools with unnatural and perverted tastes, feeble will-power, and ignorant eccentricities, strongly predisposing them to inebriety ; all they need is the exciting cause, the fertile field is prepared to receive it. The perpetuity of the race depends upon the education received, from both school and parents. It the child grows up surrounded with unnatural, morbid, influences and ignorant indulgences, and worse school education, its ruin is inevitable. Under the third group are included all those conditions and circumstances which seem to particularly favor the development of inebriety. Of these, age has a controlling influence. It is well understood that the body has various stages of development and decline. These are really physiological eras in the organization, which have prominent pathological relations, influencing the nervous system, and markedly controlling its disorders. Statistics of the time of life when inebriety begins, or when the predisposing causes are most active, are wanting. From an examination of a large number of inebriates at the Albany Penitentiary and a comparison of records elsewhere, the following facts seem to be indicated. Inebriety occurs in manhood, or midway between the weakness of infancy and the decline of old age, at a time when the brain has reached its period of fullest activity. The development of the mind and nervous system, from simple conditions to complex stages, is followed by an increase of the range of its susceptibilities and diffusion of its powers, and also greater exposure to morbid influences.

In adult life the struggle is at its height, all the faculties are intensely occupied, the fullest capacity of mental and physical energies are exercised. The ambitions, disappointments, and changes are sudden and far-reaching. The over-weaning confidence in the physical and mental powers of recovery, cause recklessness of effects. Inebriety appearing in adult life, may be the result of perversions of development, beginning in infancy. Such cases are generally associated with some form of mental disease. The brain power of infancy and youth, cannot bear much irritation without peril, its circulation is more rapid and susceptible to sudden fluctuations, which are intense in their influence. This is apparent in the cases of permanent injury, which follows from fear, anger or sudden emotions that profoundly impress the organism. At this time of life organic changes take place, which have a correlative mental transformation, the active degree of enervation existing makes the entire body extremely susceptible to commotions, in both the organic and mental developments. I believe that future observations will indicate that inebriety begins here, although its development may be delayed for years.

Sex evidently brings certain influences which bear on the causes of this affection, although the data are wanting a few outline facts indicate an intimate relation. More men are inebriates than women, but a large number of the latter are concealed by their friends; also this affection merges into other diseases more rapidly than in man.

The number of women who have mental perversions is very large, always in excess of men. Generally women suffer more keenly from inebriety than men, because they have feebler organizations, and suffer more from functional disorders and organic degenerations. Men are more exposed to the wear and tear of active life, involving accidents, excesses, wasting degenerations, with larger passions and capacities, and stronger nervous and mental organization. Women, with a less vigorous system, change oftener, and the evolution is earlier ; the absorbent system and circulation are more rapid.

In mental diseases women recover rapidly if the surroundings are good, but when functional and organic disturbances affect the organism, the return to health is slower. Both sexes at puberty, pass a period of high functional activity, a crisis in the organic and mental growth of the body, which is often the beginning of conditions resulting in inebriety. At the menapause another revolution comes over women, which predisposes to nervous defects. The close sympathy existing between the sexual organs and the encephalic circulation, predisposes both sexes to disorders of mental and functional character, of which inebriety is prominent.

Social condition has been recognized as an active cause

in inebriety ; celibacy predisposes to unnatural and perverted impulses, which rapidly develop functional diseases.
Married people lead more natural and regular lives, and
their habits and pursuits are usually fixed and methodical.
Home-life furnishes sympathy and outlets for affections
which no other conditions can give. Inebriates often
attempt marriage as a remedy to save themselves ; generally,
inebriety, beginning either before or after marriage, predisposes to celibacy. As a rule inebriates do not marry, nor
are they true to their vows after marriage. Statistics
indicate a preponderance of married men as inebriates, but
a closer examination indicates a large proportion of them
leading single lives. Domestic troubles, breaking up the
normal quietness of life, and sudden changes of conditions,
from affluence to poverty, or vice versa, often bring on this
affection by destroying the equilibrium or co-ordinating
power. Social changes frequently drive men into unhealthy conditions, which bring on perversions ending this
way. Fashions and customs develop unhealthy mentality,
which lead towards the same affection. In many instances
we can exclude every other cause but the latter, but frequently social conditions only develop and excite latent
tendencies to this affection. Future studies will show that
the seasons and weather are agents, influencing more or
less inebriety and inebriates.

The changes in the organism from these causes are
followed by corresponding changes of the mental conditions. The increased functional activity of the body in
the spring, is supposed to favor the production of mental
diseases ; but from the records of police courts, penitentiaries and asylums, inebriety appears most common and
acute in the season of autumn. This has been explained
as coming from moral and physical causes. The former
arising from general failures, with blank prospects, loss of
opportunities, and depression which follows. The latter
from physiological changes, such as thickening of the

arachnoid, shriveling of the convolutions, serous effusions, with general atrophy, and wasting of the functions. Another fact well recognized is that prolonged cold and rapid changes from one extreme to the other particularly predispose to degeneration and morbid mental and functional activity. The effect of cold, driving the blood from the surface, preventing proper contraction by congestion of the cerebral capillaries from reflex action explains this. Barometrical changes act in the same way. Sailors, soldiers, woodmen, teamsters, etc., exposed to these conditions, are generally inebriates, evidently due to similar causes.

All unhealthy mental, physical and social surroundings, continuous excitement, and dark, damp, low places of residence are noted for the inebriety which springs from them. Whatever climate or season, combined with external causes, checks or perverts the normal development of mental or physical life, becomes the exciting cause of inebriety, insanity and imbecility. There are many reasons for believing that future studies will indicate that certain climates and seasons are more favorable for the treatment of inebriety.

The temperature and condition of surroundings are evidently active causes. High degrees of heat and cold may so far impair the nutrition of the nerves, or produce paresis of the vaso-motor centres, as to result in inebriety. Sunstroke, inflammations, congestions and similar disorders of the brain-cells, coming from extremes of heat and cold, are often followed by drunkenness. A fireman while overheated, took a bath which resulted in severe headache, indigestion, and finally acute drunkenness. Here the sudden abstraction of heat so damaged the nerve-centres, as to bring on this affection. The Captain of a South American steamer informed me that the firemen and most of the men employed (for any length of time) in this service, became drunkards ; this he attributed to the extremes of climate and heat which they were constantly exposed to. Leaving

New York in mid-winter, within a few days they are in a tropical temperature, then back again, from the literal summer of the Equator to the frequent Polar temperature of New York. Inquiry among the furnace-men of Albany and Troy indicates the same fact ; men exposed to a high heat after a time (to a large extent) became inebriates.

The purity of surroundings, particularly sleeping-rooms and places of business, (without doubt) influence the develment of inebriety. Unsanitary conditions of every description react first on the nutritive functions, then on the will-power. I believe further study will show that the nutritive perversions of inebriates frequently begin in imperfect oxidation of the blood, and the inhalation of noxious impurities. In this respect our reformatories, asylums and prisons are criminally wanting. To attempt reformation or treatment in damp cellars, or rooms badly ventilated, is to ignore the predisposing causes, and fix and intensify the derangement. Cases are on record of men confined for a few months in a prison, previously temperate, who, on coming out, became furious drunkards. The New York State Prison Association mentions this fact, and the explanation is evident in the poor quality of food and bad unhealthy surroundings, strongly favoring this termination. The Almshouses and Poorhouses of the country furnish confirmatory evidence. Children and young people who have by any misfortune been confined in them, may be said to have as a rule decided tendencies to drunkenness. Older people, past the middle of life, who come to these places, go away with the same perversions. Cellar occupants and those of tenement-houses (where nearly all the conditions of health are wanting), are inebriates as a rule ; separate them from these baneful influences, and reform or recovery follows naturally in many cases. Working rooms, such as factories and shops, are often nuclei for the same causes. Sleeping and living rooms may present conditions which bring on disturbed nutrition, and then the train of causes are put to work.

The sympathy between the stomach and brain is so intimate, and the changes which take place so obscure, that inebriety may break out as the result of long and continuous abuses of the surroundings.

Mental diseases are found to be largely controlled by cosmical influences, such as Electrical phenomena, Lunar attractions, velocities and directions of winds, Geological formations, elevations above the sea level, etc., etc. A professional man of great intelligence, who is an inebriate, becomes restless, depressed, and has headache and irregularity of temper when the wind blows from the South, and a storm is impending ; also before a thunder-storm. If he can resist the cravings for liquor until the storm is over he is safe, but if the moon be at a change, and the storm is delayed, he loses all control and has a paroxysm of drinking. At the Albany penitentiary it has been observed, that chronic inebriates, recently confined, show a decided difference in work and temper on the approach of a storm.

A clergyman, who was an inebriate in early life, cannot go near the seashore, without feeling an almost irresistible desire to drink again. On two occasions he was obliged to go back suddenly to the interior to save himself—a few miles back the desire leaves him. These cases indicate a wide field of causes entirely unknown. There are influences which control the wants of the body which no one has yet solved ; conditions of exaltation and depression, which we all experience, of undefinable longings and tastes, a pressure of unknown surroundings, which may perhaps be influenced by cosmical forces. Certain districts of the far West, particularly river-bottoms, seem to present combinations of circumstances which end in this affection. The same may be said of certain kinds of labor. The State Board of Charity of New York have in their inquiries found certain rural districts where pauperism and inebriety seemed to flourish as indigenous. Who can predict what a farther study will disclose in this direction ?

CHAPTER VIII.

TRAUMATISM IN THE CAUSATION OF INEBRIETY.

From a clinical study it will be found that the use of alcohol in inebriety is in many cases only as a symptom, or one of the many causes that develop positive disease. It is proposed in the following study to show that physical traumatism is often an active cause of inebriety, which in most cases is not recognized. The early history of drinking is often a period of great obscurity, and the patient himself will have no clear idea of the conditions and causes which impel him to use spirits. His opinions are often misleading and never reliable unless confirmed by other evidence. If he has been taught to consider inebriety a vice and sin, his ideas of the early causes will be governed by this impression. If he has no fixed theories on this point, he will usually have some notion of misfortune and trouble, and consequent despair, associated with the early periods of drinking. From a clinical study the views of the patient may be of value as intimations of his present mental state, and the possible mental conditions which have obtained in the past. In all cases the tendency to exaggerate and prevaricate, without any ascertainable reason, must be considered in the problem of diagnosis. There are two distinct periods in all cases of inebriety. The first, beginning somewhere in the past, unknown and not noticeable to ordinary observers, and terminating with the first excessive use of acohol. The second, starting from this point and noted by the occasional or continuous

66

excessive use of spirits, terminating only in death or recovery. This period comes under the observation of friends and relatives, and can be accurately studied, and is supposed to include the entire field of observation. Inebriety begins in the first period, and breaks out in the latter. This first period is not studied, it is in the outer circle, the penumbra, or neurotic stage.

The second period is the umbra, and inebriate stage. In this first or neurotic stage, the causes and conditions are as varied and complex as that which produces insanity. Notwithstanding their obscurity, they often present distinct intimations of inebriety far in advance. Every case will be found to come from some special condition of change or departure, from healthy activity in the organism, in which both the function and structure are involved. Even in this early stage, a certain progressive march may be noted, often broken by long obscure halts, or precipitous strides, changing into varied forms and manifestations of disease. This neurotic stage will be marked in most cases by nerve exhaustion, instability of nerve force, and nutrient perversions and disturbances. Not unfrequently delusions and hallucinations about foods and drinks are unmistakable symptoms. Often persons who have never used spirits, and become fanatical in their efforts to reform inebriates, are in this stage, and sooner or later glide into the next one. These are the general indications, associated with innumerable minor hints and symptoms, that follow from all the degrees of inheritance, occupation, surroundings, and all conditions which make up physical and mental health. Traumatism may bring the patient into the first stage, or into the second at once. Or it may leave him susceptible to every physical state and surroundings. Psychical traumatism, or injury from mental agitation or powerful emotions, as a cause of inebriety, may be considered from two points of view. First, as a direct cause of inebriety, and second, as an indirect cause, by developing conditions which rapidly merge into this disorder. As a direct cause the following case is a good illustration :

A merchant, previously healthy and temperate, forty-five years old, with no neurotic inheritance, was returning from New York City (where he had been on business), on an evening train, on the Hudson river railroad. While moving at great speed the cars jumped the track, and ran along on the sleepers for some distance before they were stopped. The sudden alarm, crashing of the windows, and profound agitation from fear of death, produced functional paralysis, and he had to be lifted out of the car. He was taken to a farm-house, and after a few days was able to go home, but complained of exhaustion and neuralgic pains all over the body. He began to use alcohol to intoxication and could give no reason why he drank. This continued for three years, until death from pneumonia brought on by exposure while intoxicated. Notwithstanding all the efforts of himself, relatives and family, he drank precipitately to the latest moment of life. He began to drink soon after the injury, calling for it with great urgency. At first it was freely given, until he was so often under the influence that it had to be removed.

The second case of this character was that of a clergyman who was in good health, a man of strong temperance scruples, and very correct in all his habits. The sudden death of his wife from a railroad accident, threw him into a nervous fever, that lasted for two weeks, after which he began to use spirits in large quantities. He claimed that he needed it for exhaustion as a tonic, and justified his use of it to intoxication. From this period he drank at all times and places, giving no cause or reason for its use except that of a medicine. He was soon discharged from the church, and became an outcast and inebriate of the lowest grade. He is now serving out a sentence for assault in state's prison. His inebriety began directly from the shock following or caused by intense sorrow and grief.

In both of these cases there was a degree of mental and physical vigor, that gave no indications of this sequel, or any neurotic disease. There was no heredity in either case that was prominent, and the inebriety was purely from psychical traumatism.

There is another class of cases which not only have a general neurotic inheritance, but have hints of defective nerve force long before traumatism brings on inebriety. The following cases bring out these facts clearly :

A lawyer, age forty-four, who was a temperate hard-working man, was made unconscious by a stroke of lightning, and from recovery began to use large quantities of spirits at night. He became an inebriate and died three years after from delirium tremens. His grandfather on his mother's side died from inebriety, and two uncles were inebriates. His mother used spirits freely as a medicine for many years. Here it was clear that an inebriate diathesis existed, and was only developed or exploded by the traumatism.

A farmer, who was temperate, had suffered some years from nervousness and general hypochondria, was greatly excited at the burning of his barns, supposed to be the work of an enemy. He was laid up in bed for two days, then began to drink brandy, and was intoxicated from this time to death nearly every day. There was no clear history of heredity, but his nervousness and hypochondria seemed to follow from some disorder which began at puberty. Some nerve defect had lessened the vigor and integrity of the organism, and the traumatism followed, bringing out inebriety.

Another case has lately come under my care of a merchant, who had been well and temperate up to his business failure. This came upon him unexpectedly, and caused great mental anguish, follow by impulsive inebriety. He could give no reason for his drinking, and simply said it was impossible to abstain. His history, for some years

before the failure, indicated chronic dyspepsia and general
perversion of nutrition, although he had conscientiously
refrained from all use of spirits, yet his capricious nutritive
impulses exploded readily into inebriety from the action of
traumatism.

There are many reasons for supposing that all cases of
this character, where dyspepsia and nutrient disturbances
exist for some time, are peculiarly susceptible to traumatism,
particularly of psychical character. The same may be said
of a large class who have inherited unstable brain and
nerve forces, either from inebriety, insanity, or any other
organic disease. They are all more susceptible to traumatism and his results.

All these cases proceeded directly from psychical
traumatism. Whether the traumatism broke up the co-ordinating nerve centers, which are supposed to govern the
sensation of thirst, or produced some general exhaustion
of these and other centers, which found in alcohol a sedative, cannot at present be determined. The desire for
alcohol in all these cases is only a symptom of some general
nerve degeneration, which has been produced by traumatism. In the second class of cases, where psychical traumatism is the cause of inebriety indirectly, the history and
symptoms are always more or less obscure, and require
careful study. Yet these cases are undoubtedly numerous,
and will in the future attract much attention. The following cases are fair illustrations :

A banker, in middle life, in good health and strictly
temperate, was greatly shocked at the death of his father
from heart disease while at the table. For several weeks
he suffered from insomnia, and could not concentrate his
mind on anything, was nervous and complained of dull
headache. He was under treatment for a long time without any positive results. Nearly a year after he suddenly
drank to intoxication, and from this time went rapidly
down to hopeless inebriety. There was no heredity and no

ill-health up to this time. Some shock had been sustained by the nerve centers which, from the application of unknown causes, burst out into inebriety.

A lawyer whose father had suffered from general paralysis, suddenly became an inebriate at twenty-four years of age, under circumstances and surroundings that were the most adverse. After a long study it was ascertained that he had been profoundly agitated from the refusal of marriage with a lady, who soon after married a rival. That for a long time he was treated by a physician for threatened brain fever, and that he never recovered his former vigor and cheerfulness. Three years after he married into a fine family, and had every agreeable surrounding possible, when suddenly he rushed into a low saloon and drank to intoxication for the first time. He seemed to try and help himself, but every day sank lower and lower, and was finally divorced from his wife and cast adrift a hopeless incurable. The same psychical causes were at work beginning in the shock from disappointment, and slowly slumbering along until inebriety developed.

A farmer with no heredity of nerve disease, temperate from principle and a hale, vigorous man, was greatly prostrated by grief, while on a visit to the army, to find that his son had been killed in battle. For five years after he complained that he did not feel well, could not sleep soundly, was more easily exhausted than ever before, and suffered from neuralgia and changing sensations. One day, in the harvest field, he left his work and drove off to a distant city, drinking to intoxication. Ever after he drank all the time, with every opportunity and occasion. His excuse was that he had a tape worm, which had been taken into the system when he was in the army. Like the other cases the effect of grief and consequent shock produced some permanent alteration of both structure and function ending in inebriety. The cases were also noted by the absence of any stage of moderate drinking, and the sudden onset of the excessive

use of alcohol. Quite a large class after some form of
psychical traumatism have a stage of moderate drinking,
which very commonly ends in impulsive inebriety, that
comes on unexpectedly. The following are some cases,
whose origin and history are very clear in many instances :

A strong temperance advocate and lawyer of culture, in
vigorous health, was involved in a stock company that
ruined his reputation, by an accident in which he was not
in any way guilty. He suffered so keenly that he was
treated for fever, which lasted some weeks, then he resumed
business. Later, he began to use beer in moderation for
debility, and this in a year or more merged into stronger
drinks at night. Another year, and he suddenly began to
drink to intoxication every day, soon losing his business
and becoming a hopeless incurable. From the time of
failure of the company and his reputation, a steady decline
of mind and body was apparent. No special symptoms
could be recognized that pointed to other than general
failure of his former vigor and pride of character. Some
change had taken place, and something was wanting to
make up health and integrity of organism.

A second case was an engineer, with no history of heredity, and a man of fine health and thoroughly temperate.
While at his work in the field, a gun in his hands was
accidently discharged, killing an intimate friend and
brother engineer. He was greatly depressed and melancholy for months, at times would burst out into tears, and
be unable to work ; then he began to use spirits at night to
bring on sleep, and a few months after he drank to excess
and was obliged to give up his business. He died a year
later from excessive use of spirits.

A third case, equally free from all entailment of disease,
and well up to the time of great exposure and excitement
from the loss of his mill by a freshet, began to use spirits
for exhaustion and debility, and a few months later was a
pronounced inebriate, drinking all the time. It may be

asked if these cases gave indications of inebriety following the traumatism ? The answer is that the period of moderate drinking showed this tendency clearly. It could not be an accident, for all of these cases were men that were fully aware of the danger of such a course, and would not enter upon it unless impelled by a diseased impulse which they could not control.

It is exceedingly difficult for those not practically acquainted with business life to understand the constant strain and excitement which follows all business and professional activity. From the poorest laborer to the millionaire and professional man of the widest influence, there is a hurry and excitement, and a want of rest, that is steadily preparing the soil for all forms of nervous diseases. The rivalry and intensity of school life follows the child to the grave. In business, in a profession or farming, it is the same struggle for prizes ; gathering up all the energies of body and mind and concentrating them in one effort ; if they fail, turning to some other field with the same intensity and courage. A prominent professor in a leading medical college and author of note, has been a teacher of languages, a merchant, an inventor, a mining engineer, all within a career of less than half a century. These extreme changes of life and occupation strongly predispose the person to states of exhaustion, or as the late Dr. Beard wrote : " We are a nation of neuresthenics out of which many and complicated nervous diseases are constantly springing."

Psychical traumatism will appear oftener as a prominent factor in the causation of inebriety here than elsewhere. The following cases are typical of a class that represents one extreme of American life, the speculators and brokers : A broker, with no history of heredity, healthy and temperate, who had made and lost two fortunes, and was rich again, entered into a pool and sunk every dollar. His wife was taken ill, and he was forced to leave his old home, and have her taken to the hospital, where she died soon after.

He began to use spirits to great excess at once and is now a chronic inebriate. The shock from the last misfortune brought on inebriety, and although he made many efforts to recover he always drops lower from the struggle.

A banker who had made a large fortune, became a speculator and went into Wall street. He was very correct in all his habits of living and was careful of his body. For ten years he both made and lost large sums of money, and was under the usual strain of men who embark their fortunes in one venture. His son, who was expected to follow in the same business, proved a defaulter, and was sent to state's prison. The father began to drink to great excess at once, and died after three months of extreme drinking. In cases of this character there must be a condition of great exhaustion and general debility which gives way under the last shock of traumatism.

The following cases represent another extreme of life in this country. A farmer who had for twenty years worked early and late, eaten poor food, depriving himself of many necessities and comforts of life, that he might own his farm, was plunged into the deepest distress on finding that the title was wrong, and all his labors had been lost. He drank at once to excess, and a year after was taken to an insane asylum. He was discharged in a few months, but continued to drink.

A bookkeeper worked night and day in an absorbing passion for wealth, neglecting to rest and taking but little outdoor exercise. His position and investments were all swept away by a financial storm, and he became an inebriate at once. Later he was a bar-keeper, then was sent to prison for some crime. The usual explanation would be that these cases drank from despair and discouragement, but a general study will show a state of psychical pain and agony for which alcohol alone acts as a sedative. It very commonly appears in a study of cases of inebriety, that the patient will refer to some event of life, or disease, from

which he lost some power or force which he has never re-
gained. Such facts are not given as reasons for his drinking, but as explanations of his vigor or power of endurance.

One man gives a history of overwork under conditions of
great mental excitement from which he has never recovered
his former vigor. Years after he becomes an inebriate, but
he never traces the connection between the former overwork
and the inebriety. A careful inquiry will show many hints
along this interval (which may be years), that refer directly
to this event, showing that inebriety is but the result of
degenerations which begun there. In another case a man
suffers from some profound grief and sorrow, which at the
time breaks up his health, and for a long time after is felt
in general debility and weakness. Years go by, and suddenly he drinks to intoxication, and is an inebriate at once.
No good reason can be given for drinking, and possibly no
stage of moderate use of spirits precedes the inebriety. To
himself and friends a degree of ill health has been recognized from the time of his great grief, and to the physician
who can study closely this interval, there will be found
nutrient perversions, neuralgia, eccentricities, and nameless
indications of a coming storm.

A very large class of cases have in the past suffered
from some form of disease from which they have recovered
with an entailment of debility, and a want of something
that cannot be defined. They are fully conscious of diminished power, of change of vigor and force. It may be they
do not sleep as naturally, and do not get the usual rest ; or
they do not recover so quickly when exhausted, cannot
digest food as thoroughly, have dyspepsia from slight
causes. They are more sensitive than before, emotional
and excitable with every event that is irritating.

In one case a man has a severe pneumonia with a tedious,
long convalescence. After recovery, a change of disposition and character is noticed, and a year or so after he

begins to drink spirits and soon becomes an inebriate. Or
another case where a man recovers from typhoid fever, and
for a long time exhibits marked alterations of habits and
character, then suddenly or gradually becomes an inebriate.
There can be no doubt that inebriety originated in the
traumatism following the diseases in these cases. Some
special exciting cause favored its development, or possibly
the injury done to the nerve-centers would only manifest
itself in this way. The first causes are traumatic, following
the diseases or lesions which take place, particularly notable
in the complex range of psychical symptoms that are seen.
The integrity of the organism and function has been
impaired, and from this point disease and diseased tendencies are developed.

These cases are found in every community, and of
· course do not all become inebriates, but like a large class
of eccentrics are on the border line, or inner-circle shading
into inebriety or insanity. A large number of persons
engaged in the late civil war, who suffered hardship and
mal-nutrition, became inebriates years after, following the
psychical and physical traumatism received at that time.
The effects of commercial disasters, of bankruptcies, and
panics in Wall street, can be seen in inebriate or insane
asylums.

In the asylum at Binghamton, New York, for inebriates,
at one time were eighteen cases, whose inebriety could be
clearly traced to a great money panic in Wall street, known
as the "Black Friday." Many of these cases were purely
from psychical traumatism, others were already in the dark
circle close to inebriety, and needed but a slight cause to
precipitate them over. Political failures are also fertile
fields for the growth of inebriety and the action of
psychical influences. Annually a large class after the close
of a campaign find themselves literally inebriates, and if
they have money go to water-cures, inebriate asylums, or
to the far west and begin life again. The inebriety is

often of a paroxysmal or dipsomanical type, with free intervals of sobriety, that give renewed energy to the delusive hope that recovery will follow the bidding of the will. A class of moderate or occasional drinkers are always more susceptible to these influences than abstainers. This was marked in an instance where three men, two moderate drinkers and one abstainer, partners in business, with equal capital, lost it all in one night. The abstainer recovered and resumed again, the moderate drinkers both drank to excess after, and died inebriates. It may be stated as a rule that moderate drinkers suffer more frequently from psychical shocks of every form, and are more likely to become inebriates from such causes. The inebriety that follows directly or indirectly from psychical traumatism, differs in natural progress and history from other cases. The physical degenerations are more pronounced, the heart and liver take on organic disease quickly, and the mental symptoms are prominent. In some cases the course of the disease is paroxysmal, and the mental degenerations are suspicious of what is called moral insanity. As in the following : A commercial traveler in good health, and a man of character, became an inebriate dating from a steamboat accident in which he was greatly alarmed, and barely escaped being both burned and drowned. When not drinking he planned and executed deceptions, cheated his employers, and engaged in a course of crime and villainy that was without shrewdness, and entirely foreign to all his past history. He was sent to prison and died from consumption. In another case, a merchant, after the onset of inebriety from the same psychical influences, suddenly became a gambler, and frequented the lowest places of this class.

In the treatment of these cases, where the previous history has been concealed or not ascertained, the impulsive boasting and foolish prevarications, and efforts to cover up and live a double life by pretending to use all means for

recovery, and steadily thwarting them, are often clear hints
of psychical traumatism, which a more accurate history
confirms. Want of space prevents us from illustrating this
subject further. Any general study of inebriety will point
out this factor of traumatism as prominent in many cases
that are now unknown. The following conclusions may
serve as a guide to other studies in this field, or as hints of
the rich mines for clinical and psychological investigation
awaiting future discovery.

1. The injury to the nerve-centers from psychical
traumatism is the literal switch or point of departure from
the main line, from which all subsequent disease and
symptoms of change and perversion can be traced and
studied.

2. The most prominent early symptom is exhaustion or
neuraesthenia, which goes on progressively manifest in
more complex deviations from health, and general functional disturbances.

3. This may explode into inebriety at once, or appear in
moderate drinking, which will always end in inebriety.
The type of this craving will differ from others in the
extreme mental degeneration which follows.

4. The prognosis and treatment will differ materially,
depending on a knowledge of these facts, and will present
indications that are absolutely necessary to know in the
proper management of the case.

CHAPTER IX.

CAUSES CONTINUED, ADVERSITY, ETC., ETC.

The disease of inebriety has in many cases a distinct prodromic stage of moderate drinking. In a certain number of instances the disease never goes beyond this stage. Death follows from some intercurrent affection, always influenced by the state caused by alcohol. Moderate use of alcohol in a healthy man is always injurious, and is the initial stage of inebriety, whether it goes on to full development or not. All cases of inebriety are either preceded by this early stage of moderate drinking, or appear suddenly as the result of certain conditions more or less unknown. In the latter case injuries of all kinds are the most frequent causes. But in the former, the soil favorable for inebriety is already prepared, and slight exciting causes are sufficient to bring on full development.

Adversity with its mental depression and tendency to melancholy, also general lowered mental and physical activity, are among the active causes which precipitate the moderate drinker into full inebriety. Clinical experience brings abundant proof of this statement. On physiological grounds it is clear that moderate drinkers have less resisting power to disease or change, which involves exhaustion of the nerve-centers, and changes of brain force. The use of alcohol has intensified the very state of debility which it is supposed to remedy, always masking and covering up the real condition. When a special demand is made on the system, this mask is thrown off and inebriety appears.

Clinically, a class of moderate drinkers who are intently

absorbed in some pursuit or ambition, in which great interests are involved, become inebriates as a rule from adversity. They are neuraesthenic both from the use of alcohol and general perversions of the nervous system consequent upon over-work and mental anxiety. Hence, when the increased strain from the changed conditions of adversity comes on, alcohol is taken as a narcotic and gives relief.

Another class of persons are noted clinically by their weak, exhausted nerve condition, and general instability of mind and body. Adversity frequently precipitates this class into inebriety. They are always predisposed to this affection, and the use of alcohol as a stimulant is so common, and its effects are misleading, that they quickly fall into chronic stages of inebriety. The nerve-centers are in a condition of starvation from over-work and defective nutrition. Alcohol constantly increases this state, and hushes the agonized cries of nature for relief.

In the observation of every one are found cases who have been moderate drinkers up to some great trouble or adversity, then become inebriates. Another class are noticed of intensely nervous persons, who are not known as moderate drinkers, up to some period of profound shock or adversity, which changes and alters the entire mental or physical life. From this time inebriety springs up suddenly and generally progresses with great intensity. In the former case, adversity was the active exciting cause, kindling into energy conditions which may have been gathering for a long time. In the latter, it explodes a state of nerve exhaustion, in the direction of inebriety. Adversity, while depressing some men, destroying all energy and faith in any further exertion, has the opposite effect on others, rousing them to greater endeavors, which if buoyed up on alcohol, must fail in the future. The physician's frequent prescription of alcohol in cases of debility following mental trouble and sorrow, is always full of peril to the patient. At this time the danger of inebriety is very great, and cannot be over-estimated. Those who have previously used spirits in moderation, should never have it as a medicine for any general disorder marked by nerve depression.

Those who have been temperate, and suffer from great physical and psychical disorder consequent on changes of external conditions, following adversity, cannot take alcohol without great danger of inebriety.

Adversity has been recognized for a long time as preceding inebriety. The general fact has been often noted, that long-continued adversity in any community will produce more or less insanity.

The same condition of adversity is followed more frequently by inebriety, which fact can be shown in the history of every community.

Inebriety is a common sequel and physiological outgrowth of adversity. All moderate drinkers are practically inebriates in the early stages, and liable any moment to develop extreme symptoms or stages of this disease.

It may be said that the use of alcohol in all cases of nerve and brain debility, either from mental or physical depression, is dangerous and likely to bring on inebriety.

Neurasthenia is rather the comparatively permanent exhaustion which is the result of prolonged over-strain, mental or physical, or both, too little rest, insufficient or defective nourishment, long continued, until the substance of the nervous system, and often of the blood which nourishes it, is wasted or worn away far below healthy limits, entailing as a necessary consequence a corresponding *loss of nerve power* and in most cases *morbid exaltation of the sensibility*, not to speak at present of other important though auxiliary elements of such cases. Neurasthenia may be exhibited in the sphere of the mind as in weakness of thought, but especially weakness and vacillation of will, or in other words, loss of control and lack of decision ; in unduly excitable and unhealthy emotions, generally of a distressing or depressing character ; or in the sphere of the simple physiological activities, especially in feebleness and irregularity of muscular action ; exaltation of physiological sensibilities of any or all kinds, but especially *reflex* excitability ; the circulation of the blood is also, as a rule, unsteady and fluctuating, the action of the heart and small muscular vessels being easily disturbed by multitudinous causes, mental and physical. Then, again, it may be general, extending to the whole nervous system, or it may be strictly localized, involving only limited parts of the nervous system, while other parts do not participate to the same extent or at all. Again, it may be hereditary or congenital, especially the former. People are born into the world daily, to begin, live, and end life, neuresthenic, or at least to be brought early into that state ; persons who have from the beginning weak, tricky nervous systems, either as a whole, or in some of its parts. On the other hand, there are persons to be met with, at all periods in life, and of both sexes, who are naturally healthy, as the phrase goes, who from prolonged over-exertion, mental or physical, or from being under a load of care or distress, or from prolonged loss of sleep or loss of appetite, or from insufficient food, or on the contrary, excessive waste, as in hemorrhages or diarrhœas, or on account of some wasting disease, etc., and suffer a loss of balance in their nutrition, so that waste for a long time predominates over repair, to the exhaustion of the nervous energies, and a great gain in mobility, or "shakiness" of the nervous system, especially the nervous centers.

It may form a disease (as it must be called) by itself, or it may and it does form a more or less conspicuous element in a great variety of diseased states, one which behooves you all to learn and detect and estimate. It is often overlooked or under-estimated, and often is the obscure and all-important element in cases that are widely different in appearance. It is the undertone in the picture in a vast number of cases of "heart disease," "brain disease," even "softening" of the brain, of hysteria, epilepsy, melancholia, neuralgia, paresis, mental weakness, feeble circulation, insomnia, etc.   It prevails in all periods of life, and in both sexes. Instead of becoming less common, it is becoming more so, as time passes, and as people as a whole become more sedentary in habits, more intellectual in activities, more engaged by occupation or by culture, so as to augment the sensibilities at the expense of the forces or power of the nervous system.

I regard the refusal to take proper physical rest, when tired from labor, as one of the most important and powerful in inducing a love for, and an indulgence in, the use of ardent spirits. Men work till they get so tired that they cannot wait to feel sensibly rested by processes of change going on in their systems from suspension of labor. They either want to work more hours than they are able to do ; or when they have done as much as they feel themselves at liberty to do, they are so tired that they cannot rest. They get rest, therefore, in artificial ways, by resorting to eating and drinking. Some get rested by drinking tea, others by drinking coffee, others by chewing and smoking tobacco ; but the great majority of tired people in this country—and the larger share of our people are tired—drink ardent spirits in some or other of its forms or preparations. They fall back on stimulants instead of the intrinsic vitalities of their bodies. They therefore are lifted up into false conditions. Accepting these as true, they keep on working till they become so functionally impaired as to induce positive inability to work longer, or they become so constitutionally depreciated as to be smitten with incurable disease.

The hot waves which follow each other during the summer months, register their duration and intensity in the police courts, station-houses, and hospitals of all large cities by the sudden increased number of inebriates who come under observation. A sudden rise of the thermometer brings more drunken men to the station-house, and more acute intoxication is noticed on the streets. Why this is so is not clear. Why should the nerve and brain debility of inebriates seem more easily affected by extreme heat? Why should alcohol have more rapid action, causing pronounced narcotic effect? Why should the inebriate use spirits more freely at such times? These and many more inquiries await an answer from the scientists and future investigation.

One view of the subject should be practically recognized everywhere. *First*, the great danger of confining intoxicated persons arrested on the street in hot weather, in close, badly ventilated cells; such cases are in great danger of heatstroke. Narcotized with alcohol, and thrust into close, stifling air— all the favoring conditions are present, and the person is found dead next morning in the cell, or in a state of deep stupor from which he dies later. The real cause was not the intoxication, but the heatstroke from the close air of the cell. Close, hot cells should never be used for the purpose of confining intoxicated men in hot weather.

Second, in a number of cases, drinking-men suffer from partial sunstroke in the street or saloons, and are taken to station-houses, as simply drunken men. They are placed in cells, receive no care, and die. They may be temperate, and, feeling bad, take a glass of brandy for relief, fall into a state of coma, the real cause being the sun or heat-rays; but from the alcoholic breath they are judged to be intoxicated and taken to the cell, only to have an increase of their injury and die.

Another class of cases, far more common than is supposed, are those who, after a partial sunstroke, take a single glass of spirits, become delirious, and are called "crazy drunk." They are roughly taken to the station, and, perhaps, hit on the head, with no other idea than that of willfulness, and next morning are dead, or are taken to the hospital, and supposed to have meningitis, from which they die. The real cause was the policeman's club, and hemorrhage from traumatism.

Another class drink ice-water, or soda compounds, to excess, then, to relieve the distress from these drinks, take brandy or whisky and become delirious. They are arrested, and thrust into a cell like the others, and if they do not have a heatstroke suffer from injury in their delirium by striking their heads against the walls. Policemen have no other standard except the alcoholic breath for determining the state of the person.

An instance came under my observation, of a man, poorly dressed, who was overcome by heat and exhaustion, and was given a glass of whisky by a kind-hearted storekeeper. He became delirious, was taken to the station, and from thence to the hospital, where he died a few days later. The autopsy revealed a fractured skull and a ruptured artery, which came from the struggles in the arrest or self-inflicted injury in the cell.

Third, judges who administer so-called justice to these poor victims, often assume that this sudden increase of inebriates demands increased severity of punishment ; and the wrong of arresting every one indiscriminately and sending them to station-houses is still further increased. Justice is outraged, and the burdens of the tax-payer increased, and the danger to life and property made greater by recruits to the dangerous classes—classes diseased and incapable beyond recovery, yet treated as law-abiding citizens and held responsible.

The medical men in every town should insist that all men arrested during hot periods for supposed intoxication should come under medical care, and be examined carefully before they are thrust into cells. The community should be taught that the increased number of acute inebriates in hot weather points to ranges of physical causes that require study, and cannot be treated by policemen or police judges. Hot cells, in the ordinary station-houses, are

sources of danger that should be avoided. The delirious
or comatose inebriate who is placed in such cells over night,
is practically murdered. The chances of escape from heatstroke and traumatism are far less than the hope of recovery. The skill to correctly determine the condition of
these acute inebriates who are arrested in hot weather, is
far greater than in ordinary insanity, and should not be
trusted to policemen and non-experts. Here is a field for
the ambitious physician who would discover new ranges of
physical causes, and point out methods of prevention of
the greatest practical importance.

Those who use alcohol in any form, either moderately
or in excess, suffer more frequently from sunstroke and
heat apoplexies. The fatality of these cases are greater
than in those who are abstainers. The brain disturbances
following are more serious and prolonged, and in many
cases acute manias, various palsies, and dipsomaniac
impulses are very prominent.

Many of these cases occur in persons who are not
thought to be other than moderate drinkers, particularly
as they are seldom seen intoxicated, and, although at the
time of the sunstroke may have the odor of alcohol about
them, yet are not considered to be injured by spirits in any
particular way. These cases suffer from sunstroke in two
ways : either from the direct rays of the sun, or from the
heat of close rooms or areas. In the latter case it occurs
most frequently after sunset and before midnight. Very
serious mistakes are frequently made in the diagnosis by
physicians. A man will be found in the street in a state of
coma, with an alcoholic breath, which, to a superficial
observer, points to spirits as a cause. The real cause, sunstroke, is not recognized, and death follows in a cell at the
station-house, or in some other place, and nothing has been
done to avert this event. In any case of coma found in the
street in hot weather, the diagnosis of apoplexy from heat
should be considered, irrespective of all alcoholic odors in

the breath. A moderate and only an occasional user of spirits, feeling bad, took a glass of brandy, and soon after suffered from sunstroke and was taken to the station-house, where he had been ordered by a physician who made a diagnosis on the odor of his breath alone. A clergyman who drank only wine at meals, and was a red-faced man, was struck down by the sun soon after drinking some wine, and was taken to the station as a drunken man, the diagnosis being made on the same grounds. The physician should not forget that the odor of spirits, with a tendency to apoplexy, excited by sunstroke or any other cause, and the circumstance of being found insensible in the hot sun, is sufficient to warrant a diagnosis of sunstroke, rather than that of coma from alcohol.

In the second class, more difficulty follows. Thus, a man who has drank in moderation, or not at all, will be stricken down in a close room, and if the odor of alcohol is present he is supposed to be intoxicated. Cases of this kind are frequent in bar-rooms, and close, crowded tenement-houses in hot nights. The real diagnosis, heat apoplexy, is often overlooked. A man comes home from a hard day's work, takes a glass of spirits, and goes to some close room where the air does not circulate and radiation is imperfect, and soon after has an attack of heat apoplexy; or, he may go to some close bar-room, and late in the evening be stricken down.

These cases occur most frequently in large cities, and close, narrow streets, but will be found in all sections of the country where the conditions of heat and surroundings are favorable. The advice given in India to the English residents and troops is very* sensible and correct. First, to abstain from all alcoholic drinks during the hot season; and, second, to drink large quantities of water, especially in hot days. With this are many directions about the care of the body, namely: to avoid over-work, pressure of clothes, bad food, and so on. Partial sunstrokes are more common where the person has a faint, attended with dizziness, momentary loss of consciousness, followed by severe headache and great prostration. These attacks have a very serious influence on moderate or excessive drinkers. Often it is the beginning of profound degenerations, which go on rapidly to death. Heat apoplexies are very intimately associated as causes of inebriety, and when occurring in inebriates, lead to the gravest results.

The use of spirits in any form undoubtedly favors and predisposes to sunstroke ; and whatever the explanation may be, it is certain that he who uses alcohol has less vigor and resisting power to high degrees of heat.

In the comparison of the histories of many cases of inebriety, certain ranges of fact appear in a regular order. Continuous chains of cause and effect run through all the events. What appears to be the free will of the victim is but a narrow channel along which he is forced by conditions which he cannot escape. Appeals to his feelings and reason are useless, for these faculties are unable to direct or control the progress of disease. Often the victim is unconscious that he has lost his power of control, unconscious of the march of events ; the steady disintegration of brain vigor and health, and never realizes it. Delusions of health and self-control become fixed as the disease goes on. The range of his mental powers steadily narrow and approach the animal in comparison, and are finally lost in a general dissolution.

The agents exciting the emotions, and exhausting the nerve forces, and so causing inebriety, are everywhere present in the tremendous activities of our American civilization.

The atmosphere is full of psychological germs, calculated to infect the nervous system and produce disease. " Hopes and fears, appealing to the deepest motives of our nature ; political excitement, producing tumults of passion and bitter feeling ; commercial waves of good and bad fortune, causing alternately intense joy and as intense disappointment and chagrin ;" these are some of the dangers breaking up the healthy mental equilibrium and increasing the perils of every life. It is in this atmosphere that *inebriety* begins and goes to full development, standing out in lurid relief with our boasted civilization.

CHAPTER X.

Inebriety in America is more impulsive and precipitate than in other countries, the period of moderate drinking is less marked, and the average life of the inebriate is shorter. Among the many reasons for this are the tremendous activity and competition in the ordinary work of life, the intensity of living, the constant excitement and changes, filling every moment of time, calling out every energy, putting them in a constant strain, followed by want of rest, neglect of the healthy functions of the body, etc., etc. Add to this the constant practice of using the strongest alcohols at all times and occasions, and it will be seen that the average American must of necessity possess less resting power, and will fall a victim more readily to the action of alcohol.

The effects of these influences are marked psychologic-ally in the character of the inebriety seen in America. When under the influence of alcohol the average American is full of delusions of speculation for wealth, power, and political achievement ; his ideas flow in a channel of great events, and great schemes for the welfare of the nation and the race ; he is rarely a wife-beater or avenger of personal worngs, but may be prominent in a mob to destroy some great evil, or foremost to break up some old order of events which are supposed to be blocking the wheels of progress. The inebriate American will always be found in the van of every new project in politics, social science, religion, and business, and, like Colonel Sellers, is buoyed up with the stimulating hope of " There's millions in it." The records of courts rarely exhibit brutal crime among inebriates who are Americans, but great schemes of companies, frauds, and stupendous swindles, for money or notoriety, etc., etc., are common among this class. The increased consumption of all forms of alcoholic drinks, beyond the average gain of population, give indications of the increase of inebriety. This is confirmed by the court records of drunkenness in all the large cities and towns, and also from comparative evidence of the mortuary statistics which point to a rapid increase of those diseases which are most common in inebriates : namely, insanity, paralysis and acute affections of the heart, kidneys and lungs. It may be said, beyond all doubt, that inebriety in America terminates most frequently in acute organic affections of the body. Another fact of great interest is apparent to the psychological student, namely, that inebriety in America moves in waves and currents, with a decided epidemic and endemic influence. This can be traced to the rapid increase of drunkenness in towns and cities, and after a time a reaction sets in, and a marked decline follows ; the latter is seen following the temperance agitation and revivals. In some cases this is traced to special causes, such as financial depression, great social changes, etc. ; at such times moderate drinkers become pronounced inebriates and weak, nervous organizations fall into inebriety. This increases up to a certain point, then from some unknown psychical force declines, and a revival of temperance efforts follows, with a decline of inebriety to a minimum. These waves and inebriate storms that sweep over large circles of country are always followed by intense revivals of temperance interest, and are fields of the most fascinating psychological interest yet to be studied.

Alcholic inebriety among American women is undoubtedly becoming less every year. Although it is more covered up than in other countries, yet its presence is apparent in the demand for narcotics, and the sale of beer and wine by grocers, also in the divisions of saloons into general and family entrances, with separate rooms for each. Among the better classes of women wine and spirits are less openly used, and social drinking more rare. The same is true of all classes, except those of foreign birth, who still cling to the old custom of public drinking. The same general causes govern women that are noticed among men, only varying in degree ; hence, while women do not use alcoholic spirits as men do, undoubtedly they consume all forms of narcotics in excess of other classes. Their peculiar sensitive organization demands narcotics as a relief from the strain and exhaustion to which they are constantly subjected, and this is a source of great peril to the future of the American race.

In this brief outline of the nature and character of inebriety, it may be said that American inebriety is more often a pronounced form of brain and nerve degeneration, and that it comes from well marked physical conditions, largely controlled by social and psychical states peculiar to America. Its symptomology more nearly resembles that of insanity and general paralysis ; its course is in waves and currents ; its progress is shorter ; and among women the use of narcotics is more prevalent than of other forms of alcohol.

At the recent international congress, held in London under the auspices of the Society for the Study of Inebriety, brief reference was made to certain factors which contribute to make the study of inebriety in America specially serious and urgent.

Although there are abiding factors the world over, in America we have elements to study which are peculiar and unique. By America is meant the American Republic, the States and Territories bounded by the seas, the lakes and the gulf. Here sixty millions of people are placed under those physical, psychic, political and social conditions which combine to make life *more vividly intense and exacting* than anywhere else on this planet, and therefore are more susceptible to the malady of inebriism.

This region has been called "the intemperate belt." "Inebriety may be said to have been born in America and has developed sooner and far more rapidly than elsewhere; and like other nerve maladies is especially frequent here. It is for this reason, mainly, that asylums for inebriates were first organized here." Here also the total abstinence societies of modern days began. Why? because of the abnormal nerve sensibility which the feverish rush of life here has developed, a physiological condition, that will not tolerate stimulants.

Dr. Beard says that it is a greater sight than Niagara, which is presented to a European coming to this land, to behold an immense body of intelligent citizens, voluntarily and habitually abstaining from alcoholic beverages. "There is perhaps no single fact in sociology more instructive and far reaching than this; and this is but a fraction of the general and sweeping fact that the heightened sensitiveness of Americans forces them to abstain entirely, or to use in incredible and amusing moderation, not only the stronger alcoholic liquors, but the milder wines, ales, and beers, and even tea and coffee. Half my nervous patients give up coffee before I see them, and very many abandon tea. Less than a century ago, a man who could not carry many bottles of wine, was thought effeminate. Fifty years ago opium produced sleep, now the same dose keeps us awake, like coffee and tea. Susceptibility to this drug is revolutionized."

Dr. Beard makes the ability to bear stimulants a measure of nerves, and asserts that the English are of "more bottle-power than the Americans;" that it is worth an ocean-voyage to see how they can drink. A steamer seat-mate poured down, almost at a swallow, a half tumblerful of whiskey with some water added. He was a prominent minister in the Established Church, advanced in years, yet robust. He replied to the query, "How *can* you stand that?" that he had been a drinker all his life and felt no harm.

The same relative sensitiveness is shown in regard to opium, tobacco, and other narcotic poisons. The stolid Turk begins to smoke in early childhood, when seven or eight; everybody smokes, men, women, and little ones, yet the chief oculist in Constantinople says that cases of amaurosis are very few. A surgeon whom I have known, Dr. Sewny, of Aintab, after years of extensive practice in Asia Minor, has yet to see the first case of amaurosis or amblyopia due solely to tobacco. But Americans cannot imitate Turk, Hollander, and Chinese. Heart and brain, eyes, teeth, muscle, and nerve are ruined by these vices, yet the frightful fact remains that laterally the importation of opium has increased 500 per cent.! The "tobacco heart" and other fatal effects of cigarette smoking are attracting the attention of legislators as well as physicians, and the giving or selling this diminutive demon to youth is made in some places a punishable offense.

Physical, psychic, political, and social conditions combine in the evolution of this phenomenal susceptibility. Nowhere, for instance, are such *extremes in thermal changes*. I have seen in New England a range of 125°, from 25° below to 100° above, in the shade. The year's record at Minnesota has read from 39° below to 99° above, a range of 138°. Even within twenty-four hours, and in balmy regions like Florida, the glass has shown a leap from torrid heat to frosty chill.

No wonder then the greatest fear of some is the *atmosphere!* They dreaded to go out to face Arctic rigor or tropic fire, and so get in the way of staying indoors even in exquisite weather of June and October. They make rooms small, put on double windows, with list on the doors, and build a roaring furnace fire in the cellar, adding another of bright anthracite in the grate. The difference be'ween this hot, dry, baked air within, and the wintry air without, is sometimes 80°. It is estimated that the difference of temperature inside and outside an English home averages 20°, and that within and without an American dwelling is 60°. The relation of this to the nervousness of the people is apparent.

The uniform brightness of American skies favors evaporation. The Yankee is not plump and ruddy like his moist, solid British brother, but lean, angular, wiry, with a dry, electrical skin. He lights the gas with his fingers, and foretells, with certainly, the coming storm by his neuralgic bones. Hourly observations were conducted for five years with Captain Catlin, U. S. A., a sufferer from traumatic neuralgia in care of Dr. Mitchell. The relation of these prognostic pains to barometric depression and the earth's magnetism was certified beyond doubt, and was reported to the National Academy of Science, April, 1879. Even animals in the Sacramento valley and on the Pacific coast are unwontedly irritable while the north desert winds are blowing, and electricity seeking equilibrium, going to and from the earth. Fruits, foliage, and grass, towards the wind, shrivel. Jets of lightning appear on the rocks and sometimes on one's walking stick.

But *psychic and social factors* cannot be ignored. Someone has said that insanity is the price we pay for civilization. Barbarians are not nervous. They may say with the Duchess of Marlborough that they were born before nerves were invented. They take no thought of the morrow. Market returns and stock quotations are unknown ; telephones and telegraphs ; daily newspapers, with all their crowded columns of horror and crimes, are not thrust upon them ; and the shriek of the steam engine does not disturb their midday or their midnight sleep. Once a day they may look at the sun, but they never carry watches. This bad habit of carrying watches is rebuked by a distinguished alienist, who says that a look at one's watch, when an appointment is near, sensibly accelerates the heart's action and is correlated to a definite loss of nervous energy. Every advance of refinement brings conflict and conquest that are to be paid for in blood and nerve and life.

Now, it is true, that watches are occasionally seen in England. Sun-dials are not in common use in Germany and Switzerland. But the "American Watch " is an institution. Not the Elgin, the Waterbury, or any particular watch, but the worry and haste and incessant strain to accomplish much in a little time—all this symbolized in the pocket time-piece, is peculiarly American. It was an American who, at Buffalo, I think, wanted to wire on to Washington. When told it would take ten minutes, he turned away and said, "I can't wait." He now uses the Edison telephone, and talks mouth to mouth with his friend. Dr. Talmage says, "We were born in a hurry, live in a hurry, die in a hurry, and are driven to Greenwood on a trot !" The little child, instead of quietly saying to its playmate "Come," nervously shouts, "Hurry up !" You cannot approach the door of a street car, or railway carriage, but what you hear the same fidgety cry, "Step lively !"

Said a New Yorker to me, " I am growing old five years every year." Can such physical bankrupts, whose brains are on the brink of collapse, bear the added excitement of drink ? The gifted Bayard Taylor was but one of thousands who burned a noble brain to ashes in a too eager race of life. Reviewing sixteen months he notes the erection of a dwelling-house, with all its multitudinous cares, the issuing of two volumes of his writings, the preparation of forty-eight articles for periodicals, the delivery of 250 lectures, one every other day, and 30,000 miles travel. The same story might be told of other brain-workers who never accepted the "gospel of rest."

The *emulous rivalries of business life* and the speculative character of its venture cannot be paralleled elsewhere. The incessant strain they impose increases mental instability.

Bulls and bears, pools, corners, margins, syndicates, and other "words that are dark, and tricks that are vain," represent the omnivorous passion for gambling. Millions may be made or lost in a day. No one is surprised if a Wall Street panic is followed by suicides.

Legitimate business may, by its methods, exert a pernicious influence on the nervous system in still other ways, as for example, in the depressing influence from specialization of nerve function, as indicated by Dr. Jewell, where one keeps doing one petty thing monotonously year after year and so sterilizes mind and muscle in every other direction.

Turning to *educational systems* in America, we see how unphysiological they are, and calculated to exhaust the nervous energy of youth, many of whom have inherited a morbid neurotic diathesis. Of twenty-seven cases of chorea reported by Dr. Hammond of Bellevue Hospital, eight were "induced by intense study at school." Dr. Treichler's investigations as to "Habitual Headache in Children," cover a wide field and show that continental communities suffer from similar neglect of natural laws. Here it is more notorious.

Not to dwell on these points, we may say that the *stimulus of liberty* is a productive cause of neurasthenia in America. It is stated that insanity has increased in Italy since there has been civil and religious liberty guaranteed. A *post hoc* is not always a *propter hoc.* But it is obvious that the sense of responsibility which citizenship brings ; the ambitions awakened by the prospect of office, position, power, and influence ; the friction and disquiet, bickerings and wranglings, disappointment and chagrin that attend the struggles and agitations of political life, do exhaust men, and more in a land where opportunities for advancement are abundant as in America. While writing these words, news is received of the sudden death of a prominent New York politician, comparatively young, directly traceable to disappointment in carrying out a scheme on which his heart

was set.   Chagrin acted like a virulent poison on a system
already unstrung by the severe political struggle in which
he was defeated.

Multitudes contract the full malady of inebriism under
the continued pressure of these political campaigns.   The
patient of a friend of mine had, for two years, been kept in
working order.   He was living, however, on a small reserve
of nerve force.   A few days before election, he was drawn
into a five minutes eager discussion, and became entirely
prostrated, more exhausted than by months of steady work.

Other nations have their measure of liberty and aspirations for social and political eminence to gratify.   But
nowhere have men the exhilarating possibilities of position,
wealth, and influence, that this republican community offers.
The history of the last half century, as related to this fact,
reads like a romance.   But liberty, like beauty, is a perilous
possession, and it has been truly said "the experiment
attempted on this continent of making every man, every
child, every woman an expert in politics and theology is one
of the costliest of experiments with living human beings,
and has been drawing on our surplus energies for one hundred years."

Finally, *American life is cosmopolitan.*   A curious observer
noted nine nationalities in a single street car in New York,
one day.   I repeated the fact to a few of my students who
were riding with me through those same streets.   Looking
over the ten or dozen passengers on board, one of them at
once replied, " Well, here are *five* nationalities represented
here."

In one aspect, these importations, particularly English,
German, and Scandinavian, are compensative and antidotal.
We may hope, with the author before quoted, that "the
typical American of the highest type will, in the near future,
be a union of the coarse and fine organizations ; the solidity
of the German, the fire of the Saxon, the delicacy of the
American, flowing together as one ; sensitive, impressible,

readily affected through all the avenues of influence, but trained and held by a will of steel ; original, idiosyncratic ; with more wiriness than excess of strength, and achieving his purpose not so much through the amount of his force as in the wisdom and economy of its use."

This hope may be realized in the future and in the highest type of American manhood. It is a bright, optimistic view of things, but we have to do with the present and the evils of society as they exist. We have to face the fact that our civic life is growing at the expense of the rural ; that our cities are massing people by the hundreds of thousands, among whom, on the grounds of contiguity, association, and psychic sympathy, evil influences become more potent to undermine the welfare of society ; that we have to encounter in America the drink traffic in its belligerant aspects, as nowhere else, not only politically and financially organized most thoroughly, but ready it would seem to use fraud, violence, or assassination if other means fail, and that we have anarchism stirring up discontent and firing the passions of the desperate classes, who understand liberty to mean license, equality to be the abolition of all the diversities of position and property which intelligence, temperance, and industry have made, and will make, to the end of time.

We have had a practically unrestricted importation of the refuse population of Europe. Of every 250 emigrants one is insane, while but one of 662 natives is insane. Add to all these facts the conditions of American life already enumerated as related to the development of neuroses, particularly inebriety, and we have material which makes the study, as was stated at the start, serious and urgent.

CHAPTER XI.

PREVALENCE OF INEBRIETY AND ITS MORTALITY, ETC.

Notwithstanding all the advances of civilization and intelligence, and the increasing temperance agitation and effort, the drink evil, or inebriety, is one of the most threatening, ominous perils to all social progress and development of to-day.

Some idea of its extent may be obtained from the following satistics. In 1891 over five hundred thousand persons were arrested in this country charged with being drunk and disorderly, crazed from the effects of alcohol. Nearly twenty per cent. had committed crime of petty character while under influence of spirits.

These facts furnish some approximate estimate of the extent of inebriety in this country. It is reasonable to infer that if this number of inebriates have come under police surveillance in one year, at least a third more using alcohol to excess, have escaped legal notice.

The statement that there are fully a million persons in this country who are, continuously or at intervals, using alcohol and opium to excess and poisoning themselves, is not an extravagant or improbable one.

Of course the mortality from this, both directly and indirectly, is excessive, and the oft repeated assertion of sixty and a hundred thousand deaths a year from alcohol is, after all, only a minimum statement.

The physical and mental degeneration, the unsanitary conditions of living and surroundings, predisposing to pauperism and exhaustion, all strongly favor disease, accident and early death. Dr. Kerr, of London, after a very exhaustive study of the mortality reports of the United Kingdom, concluded that the annual mortality from inebriety could not be less than one hundred and twenty thousand.

This statement was confirmed by Mr. Wakely, a well known coroner, who found in fifteen hundred inquests held in one year, nine hundred of them were from inebriety. The conclusions of the late Mr. Neilson, the eminent actuary, are equally striking. He found that while ten temperate persons from 15 to 20 years of age die, eighteen inebriates of the same age die. Between 21 and 23 years (inclusively) 51 of the intemperate of inebriates die for every 10 of the temperate. Between 31 and 40 about 40 die. In the first group the fatality is raised 80 per cent., in the second over 500 per cent., and in the third 400 per cent. Mr. Neilson calculated the chances of life as follows : A temperate adults chances of life is at 20 over 44 years, an intemperate 15½ years ; at 30 a temperate 36½, an inebriate 13.8 years ; at 40 a temperate's 28.8 years and an inebriates 11.6.

In the mortality of Great Britain only between 1400 and 1500 deaths are noted as arising from alcohol annually. But it is well known that these figures are no criterion of the actual number of deaths from inebriety. The present system of death certificates makes it easy to evade the real cause of death, and put down some intercurrent affection or condition that occurs just before death takes place. This is done to save the feelings of friends and relatives, and cover up some state which would only add new pangs of sorrow by exposure.

A new system of registration is demanded to ascertain accurately many of such mortality facts.

The Harvian Society of London have studied this topic and reached similar conclusions, sustaining Dr. Kerr's estimate of one hundred and twenty thousand deaths yearly in the United Kingdom from inebriety. In this country the late Dr. Parker showed from some comparative facts that inebriates had an average of ten years, after the full devel-

opment of the drink curse. Dr. Harris indicated that twenty
per cent. of all deaths in large cities were due to inebriety.

All the life insurance companies find the mortality which
dates from heart and kidney disease associated with the
use of alcoholics, and many of them refuse to take risks on
any one but total abstainers. It is computed that for every
death from inebriety there are at least fifty cases of illness.
Associated with this are premature deaths, starvation, accidents, violence, neglect, until the loss and suffering becomes
appalling in magnitude and severity.

The recent outbreak of cholera, with its extreme fatality,
recalls the mortality among inebriates which has characterized the march of this and other epidemics in modern times.
We select illustrative statistics of the cholera epidemic
of 1832. In St. Petersburg, out of 10,000 deaths only 145
were known to be temperate ; in Moscow, only 2 out of 6,000
cases were temperate. This fact so alarmed the citizens that
nearly all the population ceased to use alcohol ; of 30,000
victims in Paris, nearly every one used alcohol, in some form,
to excess ; nine-tenths of those who died in Poland, were of
this class. In some towns *every* inebriate was swept away.
In Tifflis, *every* drunkard died. In the Park Hospital of
New York city, only 4 persons were temperate in 200 fatal
cases. In Albany, there were only 7 out of 326 fatal cases
who were not inebriates. In the late epidemic of yellow-
fever in the South, the percentage of victims among inebriates was nearly as large. These are not extraordinary
facts, but follow, naturally, the degeneration produced by
alcohol, and are readily explained by the low vitality and
lessened power of resistance to toxic forces and agents
present in every inebriate. Most unfortunately, this condition is not realized by either the patient or friends until
it is too late. The continued use of alcohol keeps up the
delusion of strength and vigor ; but with the onset of disease all is thrown off, and only the physician and surgeon
can realize their hopeless condition.

CHAPTER XII.

Any one who studiously watches the evolution and dissolution of families, some of whose members are addicted
to alcoholic excess, must be struck with the frequent
occurrence of pulmonary phthisis among them. So on the
other hand, it is no less astonishing to find the latter
disease suddenly appearing in families who are absolutely
free from a phthisical history, and who seemingly live
amidst the most healthful surroundings. Why these two
conditions should be so closely associated, if in consonance
with the current belief, the one is a nervous, and the
other a strictly pulmonary disease, is not very clear. The
following is an attempt to an elucidation of this intric-
ate problem, in which I shall endeavor to show that
these two apparently isolated phenomena are naturally
interchangeable with each other, and that like two diversi-
fied islands cropping out above the surface of the ocean
without exposing their connection beneath, they find their
common bond of union in a disordered state of the nervous
system.

In order to make this subject as practical and as intel-
ligible as possible, I shall at the very outset endeavor to
prove the intimate association between alcoholism and
phthisis, how one link may change place with the other in
the chain of vital persistence, by citing a number of living,
illustrative examples. The first ten of these cases have
been culled from the extensive experience of Dr. Crothers,

and have been placed at my disposal through his kindness ;
while the remainder have been obtained from various other
sources.

Case I. J. B., aged 42 years, began the excessive use of
spirits after the death of his wife. He was a merchant,
temperate, prosperous, and a man of character. He became
a steady drinker, and was practically intoxicated all the
time. After an attack of delirium tremens he was placed
under my care. During the four months while under
treatment, he was alternately depressed and elated. He
complained of wandering pains, and changeable appetite,
as well as of spasmodic periods of coughing. A few
months after he left me, he relapsed and continued to
drink until he died a year later.

His mother and two sisters died of consumption. His
father died from injury, but his grandfather was asthmatic,
and used spirits to excess for years. One uncle on his
father's side died from excess of drink, and another one
died of consumption. One uncle died from phthisis after
many years of drink excess.

His grandfather on his mother's side drank more or less
all his life, and died from some rheumatic trouble.

Case II. B. A., aged 35, a mechanic, began to use spirits for
insomnia and general debility, and finally became a periodical inebriate. He was under treatment for six months, and
recovered. His father, grandfather, and two uncles, died
of consumption. His mother was hysterical, and his grandmother on his mother's side died of some lung trouble. One
brother died from chronic alcoholism, and a sister is a drug-
taker.

Case III. C. H., age 48, an army officer, began to drink
during the late war. He is now a dipsomaniac, with distinct free intervals of three months. His mother died of
consumption two months after his birth, and his two sisters
died of the same disease. His father's family is temperate,
but several members have died of consumption. His grand-

father on his mother's side was a sailor, and drank to excess at times.

Case IV. D. P., age 38, a farmer. His drinking seems to date from a nervous shock following the burning of a barn by lightning. His two brothers are chronic inebriates, one sister is a morphine taker, and the other uses both spirits and drugs to excess for all kinds of imaginary evils. On his mother's side, a grandfather and three aunts and one uncle died of consumption. His mother is still living. His father died of pneumonia, and his grandmother on his father's side died of consumption.

Case V. E. J., age 31, a clerk of inferior mental and physical development, began to drink at puberty. Consumption has been the common family disease of both parents. On his mother's side both consumption and inebriety have been common. On his father's side consumption alone has prevailed.

Case VI. P. O., age 28, is without business, and drank from infancy. He is now a chronic inebriate and has had delirium tremens. His father and two uncles died of consumption. His mother is a woman of wealth and fashion, and she lost her mother and one sister from consumption.

Case VII. M. B., a lawyer, 54 years old, who began to drink at fifty from no apparent cause. His father and grandfather died of consumption at fifty years of age.

Case VIII. D. T., age 38, a conductor, began to drink after an injury to the spine. A brother, who was injured at the same time, died of consumption. The mother and a sister, the grandfather, and grandmother, on his father's side, all died of consumption.

Case IX. D. B., 24 years old, and without business, began to drink at puberty, and is now a chronic inebriate. Both parents died of consumption. His grandfather on his father's side, and two uncles on his mother's side, died of the same disease.

Case X. A. H., 34 years old, a physician, took morphia

for malarial poisoning, and then used alcohol to great
excess. His mother and three uncles on his father's side
died of consumption. His older brother became an inebriate at about thirty years of age, and one sister is in Colo-
rado to prevent consumption.

Case XI. (*Quarterly Journal of Inebriety*, Oct., 1888, p. 390.)
"George Ulmer came from England in 1798 and settled at
New Haven, Conn. He was a harness-maker, a beer-
drinker, and after middle life drank rum to excess, until
death at sixty-one years of age. His wife was a healthy
woman, and lived to eighty years of age. Eight sons grew
to manhood and married. Six of them died of consumption
under forty-five years of age. One was killed by an
accident, and one died from excessive use of spirits. Two
daughters grew up and married ; one died of consumption,
the other in childbirth. They left four children ; two were
inebriates, and the others were eccentric and died of consumption. Of the children of the eight sons only ten grew
up to manhood. Four of these drank to excess and died.
Three of the six remaining died of consumption, and two
others were nervous invalids, until death in middle life.
The last one, a physician of eminence, has become an
inebriate and is under care at present. He is the only sur-
viving member of all this family. The male members of
this family were farmers, tradesmen, and men of more than
average vigor in appearance. They married women (so far
as can be ascertained) without any special hereditary history of consumption or inebriety."

Case XII. Father was an inebriate until after he was
forty years old, at which time a cardiac affection developed
itself from which he ultimately died, but which had the
power of restraining him from exercising his morbid appetite. His brother was a drunkard too. Three of his sons
became confirmed alcoholics, one daughter died of phthisis,
and another son died of general paralysis.

Case XIII. Father violent, an alcoholic, and **a libertine.**

Mother is very nervous, and died of cancer of the uterus. Many of patient's relations are drunkards. Her brother and sister died of chest disease, and another brother is always ill, and coughs a great deal. She was admitted May 3, 1879. One month previously she had a chill, rigors, and feverishness, which confined her to bed for four days ; then she began to cough, and had two copious hæmoptyses. She sweats profusely at night, is losing flesh, and in a word has all the symptoms of pulmonary phthisis. Physical examination shows evidence of tuberculosis of both apices.

The histories of these cases give the most unmistakable proof that alcoholism and phthisis are not mere coincidences, but that they have a relationship so intimate that one may be converted into the other. The problem arises, however, as to the channel through which alcohol produces phthisis ; for if these two conditions are interchangeable, it is obvious that they must possess a common physiological basis, and this I believe resides in the nervous system. I have elsewhere (to which I beg to refer the reader) given good reason for believing that pulmonary phthisis is principally nervous in character, and by considering it as such, the natural association between the two diseases is at once established. For whatever else may be said of the action of alcohol, it is pretty generally understood that it possesses a special affinity for the nervous system, and that it produces its principal ravages in the body by operating on this, and by preference on the peripheral nervous tissue. Dr. James Jackson, in this country, and Dr. Wilks, in England, were I believe the first to point out this form of disease, and they called it alcoholic paralysis. It has since then received the more appropriate name of alcoholic neuritis, and it is characterized in its early stages by numbness, tingling, hyperæsthesia in the extremities, and later on by anæsthesia, paralysis of motion, loss of knee jerk, quickened pulse, shortness of breath, and frequently by pulmonary embarrassment. The brain and spinal cord remain comparatively

normal. The morbid changes occurring in the peripheral
nerves under the influence of alcohol are parenchymatous
and interstitial, or in other words they are confined to the
nerve substance itself, or to its investing membrane. As a
rule these changes occur together, the latter in many
instances depending on the former, but frequently one
exists exclusively of the other ; especially is this true of the
degeneration of the nerve fibre itself.

It being established, then, that alcohol has the power of
producing degeneration of the nerve fibres, it does not
require a reckless flight of fancy to see how, by operating
on the same tissue, it may bring about that peculiar
destruction of lung substance known as pulmonary consumption. Degeneration of a nerve implies degeneration
of the organ which it supplies with sensation and motion.
Thus, degeneration of the sciatic nerve is followed by
impairment of sensation and motion in the muscles and
other textures of the leg—a condition which is almost constantly present in chronic alcoholism, and degeneration of
the pneumogastric nerves is just as naturally followed by
disease of the lungs, heart, stomach, and all the other
organs supplied by them. This is no more than we may
legitimately anticipate ; for it has been amply proven that
division of, and protracted pressure of tumors, aneurisms,
etc., on the pneumogastric nerves are capable of calling
forth all the destructive lesions of pulmonary phthisis.

The following cases will serve to illustrate the close
anatomical and physiological association of chronic alcoholism and phthisis, as well as other destructive pulmonary
charges with degeneration of the vagii, and of the respiratory center (the latter of which practically amounts to the
same thing), and with that of the peripheral nerves. The
difficulty encountered in this research has not been so much
in obtaining an abundance of material in which phthisis
was evidently the direct result of alcoholic abuse, as it has
been in finding the records of cases possessing all the points

which I desire to emphasize in this paper, viz.: the coexistence of pulmonary disintegration, alcoholism, and nerve degeneration, well brought out by a thorough *post mortem* demonstration.

Case XIV. Drs. Oppenheim and Siemerling. Male, addicted to alcoholic excess, was received in hospital Jan. 26, 1886. He was weak and stiff, but had no pain. At the end of the same month he became delirious, and also paretic in both legs and arms. Death occurred in March of the same year. On section it was shown that the heart was normal, and that he had pneumonia ; microscopically it was proven that the radial, peroneus, and saphenous nerve had undergone degeneration. Not stated whether the vagii were examined or not.

Case XV. Drs. Oppenheim and Siemerling. A female, age 45 years, suffering from chronic alcoholism, was received Dec. 26th, and died on the 28th of the same month, in the year 1885. On section there was found chronic exudative pleurisy on right side, as well as a caseous bronchopneumonia and tracheitis. The great saphenous and superficial peroneus nerves had undergone parenchymatous degeneration of a medium degree. No other nerves were examined.

Case XVI. Dr. T. Déjerine. Female, age 46, a hard drinker, suffered from paralysis of both upper and lower extremities. Had a pulse of 150–160, and her heart sounds were normal. Her death was caused by pneumonia. Section showed parenchymatous neuritis of the cutaneous and muscular nerves, as well as of both vagii in the cervical region.

Case XVII. Prof. Schultze. Male, 39 years old, developed diabetes insipidus in 1882, but had been feeble since childhood. He used alcohol greatly to excess in his younger days. Some time after the year 1882, he began to suffer from nystagmus, trembling in the arms, perversion of sensation (paraesthesia) in the legs, and from thoracic constriction. In 1886 he became subject to marked attacks of dyspnoea, and death was caused by paralysis of respiration. Section : Degeneration of medulla oblongata and spinal cord, as well as that of the root of the vagus and hypoglossus. No account of the *post mortem* appearances of the lungs is given, but it is evident that these organs were implicated in the morbid processes, since death was produced through pulmonary paresis.

Case XVIII. Strümpbell. Male, aged 47 year, a potator, was received November 25, 1881. His frame is large and powerful. Both of his arms hang helplessly by his side ; hands œdematous, skin and tendon reflexes wanting ; legs weak and powerless ; pulse, 124 ; temp. 38.2° ; deglutition and power of speech impaired ; after a while œdema of lower extremities, cough, diarrhœa, dyspnoea ; bronchial râles, paralysis of diaphragm, and death, Feb. 13, 1882. Section : Marked tubercular phthisis of both lungs. The radial median, crural, and sciatic nerves were degenerated very decidedly, and Dr. Strümpbell believes that the phrenic and vagii were also involved, but he failed to examine them closely.

Case XIX. Drs. Oppenheim and Siemerling. Male, 26 years old, a potator, was received in the Charity Hospital Jan. 17, 1881, on account of delirium tremens. He complained of headache, giddiness, and formication in the legs. He improved and was dismissed, but was received again on July 28, 1883, on account of marked disturbances in the nervous system. He now suffered from complete impotence, lancinating pains and rectal tenesmus. In August, he became subject to polydipsia and polyuria ; on the 12th of December, there was dullness in left supra clavicular fossa, and infiltration of both apices and tubercle bacilli were found in the sputum. He gradually sank and died in August. Microscopic examination showed degeneration of the medulla oblongata, and of all the peripheral nerves which were examined.

Case XX. Dr. Oswald Vierordt. Male, 30 years old, much addicted to alcohol, and without a syphilitic history, suffered since March, 1884, with piercing, lightning pains in the lower extremities, as well as with weakness, unsteadiness, and formication in the same. He also developed tubercular phthisis and died the following March. Section : extended tuberculosis of the lungs and degeneration of the columns of Goll, medulla oblongata, and the cervical and dorsal portions of the spinal cord.

Case XXI. Mr. Sharkey (*British Medical Journal*, 1888, April 21, and *Journal of Inebriety*, Jan., 1889, p. 67) related a case of alcoholic paralysis of the phrenic, pneumogastric and other peripheral nerves before, and presented specimens of the same to the Pathological Society of London. The patient was a female and addicted to the excessive use of alcohol. She suffered from weakness in her legs, numbness and cramps, and was incoherent in speech. Respiratory sounds were harsh, and in a few days after admission had a rigor, which was followed by a temperature of 102.8°, severe attacks of dyspnoea, paralysis of the diaphragm, and difficulty in swallowing. Respiration 40 per minute, and average pulse rate 140. Spitting of blood supervened, the lung apices began to break down, and she died after having been under observation nearly one month. Section : tuberculosis of both apices and inflammatory changes in the phrenic, pneumogastric, and popliteal nerves.

In these examples we have proof that pulmonary phthisis can be produced through the toxic action of alcohol on the nervous system. This is unquestionable in four of the cases, and in so far as demonstrating the mode of the action of alcohol on the human lungs is concerned, it is equally true of the other cases ; for I think it is pretty well established that phthisis is but the legitimate offspring of any persistent catarrhal state of the lungs, and that chronic bronchitis and catarrhal and broncho-pneumonia, are but the milestones marking the pathway along which the disease travels to its final destination.

Such then being the relation between alcoholism and pulmonary phthisis it is very readily understood why these two diseases should so frequently change places in different members or generations of the same family, and why they are so often associated with various other nervous disorders. Moreover, alcohol having the potency to produce phthisis *de novo* in the human subject, either directly or through hereditary influence, or both, as we have seen, it must, in view of its past and present widespread abuse in civilized countries, be a tremendous factor in maintaining the ranks of the hundreds of thousands of those who are annually slain by this terrible malady. To this and to no other conclusion do the premises of this paper point, and if one had the inclination to moralize on this subject it would be very interesting to inquire why the North American Indian, and other savages, were practically free from pulmonary consumption until they came in contact with the white race ! When we connect the facts that alcohol and syphilis are the greatest curses which the Indian has acquired from his white civilizer, with the evidence which has been brought forward in this as well as in another paper on *Syphilitic Phthisis*, I think it must be obvious that these two causes are largely responsible for sowing the seeds of pulmonary phthisis among these people.

CHAPTER XIII.

EFFECTS OF ALCOHOL AND BEER ON MENTAL FUNCTIONS AND THE BRAIN.

The effects of a single dose of alcohol differ widely in different individuals, and this lies at the root of all scientific inquiries into the matter. The variety of the effects on the mental faculties of different brains is also extreme. This indicates such different qualities and susceptibilities in different brains as regards this agent, that it makes the whole question of the effects of alcohol a most complicated one, not to be explained by a few unqualified assertions. In reply to the question, What are the normal effects of alcohol on the mental forces of the brain ? the scientific man must reply, What kind of brain do you mean ? and it is only by a careful study of the qualities, the tendencies, and potentialities of different brains, that we can answer the first question properly. We need to study the mental qualities of the brain at different periods of life, in the two sexes, in different temperaments and constitutions, in different races, in different states of health and vigor, and with reference to the hereditary tendencies of the organ ; for all these things influence the effects of one single small dose of alcohol. So we find, looking from the point of view of the amount of the doses, the effect is very different. There is, I believe, no other agent known which differs so greatly in different instances in the dose needed to produce the same effect on the mental powers as a dose of alcohol, and herein again we find that there must be the greatest difference in the power of resisting the effects of alcohol in different brains. Taking the lower animals, that difference is exceedingly small, an ounce of alcohol given to a dozen dogs of the same size will practically have the same effect on them all ; but an ounce given each to a dozen men has not only the most different effects in the mental faculties it stimulates, as we have seen, but in the *amount* of the effect it causes. Some brains are exceedingly sensitive to very small quantities ; other brains have the power of resisting or tolerating alcohol in a wondrous degree, this being an innate quality quite apart from the effect of the use and custom. These differences are so great as to compel us to conclude that there are enormous inherent disparities in human beings in this respect, and this is no doubt one of the very great dangers in the use of alcohol.

So we also find at the various periods of life, ordinary small doses of alcohol have very different effects. In a child the effect is extremely great, in a boy or girl it is also great, but it is not so great in a growing adolescent. In the two sexes there are also considerable differences, the female having less resisting power, her brain being unusually much more susceptible to the influence of this agent. Looking at different races, the difference of effect of the same dose is also extremely great There are some savage races that are so subject to its influence that a very small dose indeed—half an ounce—will have greater effect on them, than two or three ounces will have on an ordinary European. The psychological, the mental effects of small doses of alcohol are therefore exceedingly various, and we have not yet discovered the precise qualities of brain which caused these differences. We can not tell beforehand which brain will be susceptible to its effects, and which will not. Looking at the matter next from a point of view of the effects of a much larger dose, these will be found much more uniform. The effect, instead of being stimulating, is then narcotic, and we have a deadening, paralyzing and temporary arrestment of the mental functions of the brain in every individual if a sufficient quantity is taken.   But here we find much variety in the way the result is arrived at, when carefully studied.

In one person we have this paralysis, this deadening taking place first on the intellectual faculties, in another on the emotional, in another on the propensities, and in another on the power of motion.   We see a certain kind of mental degeneration of a slight type, which results in those who habitually take an amount of alcohol that is to them excessive.   This slow but quite marked type of mental degeneration a doctor of experience soon comes to observe in his patients ; and others a certain change mentally, morally, and bodily, in the man who is taking more than is good for him.   The expression of his face and eyes— those mirrors of the mind—you see has changed, and for the worst.   The mental condition of the man is lowered all round, and especially one effect is noticed, that his higher power of control is lessened.   I am safe in saying that no man indulges for ten years in more alcohol than is really good for him, without this kind of degeneration being observed, and that although during these ten years he was never once drunk, we find him psychologically changed for the worse in his independence of mind, in his spontaneity. After a man has passed forty, such changes are very apt to be faster, and more decided.   We see such a man's work and his fortune suffering, but we dare not call him either a drunkard or dissipated, because, as a matter of fact he has never been drunk, and never intends to be drunk.   Whether this degeneration takes place soon or late depends upon inherent resistive capacities of his brain cells.   In some individuals the resistive capacity against alcohol is so great that for years they may indulge in its excessive use without this degeneration taking place to any great extent, but in other instances we have it very rapidly developed indeed.

Some men pass into a premature old age and become old at fifty, when they ought to have lived on and been young

men up to sixty, and this merely owing to the excessive use
of alcohol. Memory and the power of thinking are affected,
but you see the lowering most in the finer faculties, the
tastes, the more delicate perceptions of things, and the force
of character. This is an effect which, I believe, is especially to be observed in men who have used their intellectual
powers constantly and vigorously. We often see this effect
on the brains of men in our profession, of medicine, at the
bar, and even among the clerical profession, in a very marked
degree, without their owners having been once drunk. In
such persons, their mental powers having been greater to
begin with, and with a finer edge on them, you notice in a
more marked way this degeneration in its progress. This,
I may say, is the least marked mental effect of alcohol
taken, not so as to produce drunkenness, but taken in
greater quantity than the physical constitution of the brain
can stand over a long period. In some brains a very small
quantity indeed, taken daily, will produce this degeneration.

For some years past a decided inclination has been
apparent all over this country, to give up the use of whisky
and other strong alcohols, using as a substitute, beer
and bitters and other compounds. This is evidently
founded on the idea that beer is not harmful and contains
a large amount of nutriment ; also that bitters may have
some medicinal quality, which will neutralize the alcohol it
conceals, etc. These theories are without confirmation in
the observations of physicians and chemists where either has
been used for any length of time. The constant use of beer
is found to produce a species of degeneration of all the
organism, profound and deceptive. Fatty deposits, diminished circulation, conditions of congestion, and perversion
of functional activities, local inflammations of both the liver
and kidneys, are constantly present. Intellectually, a stupor
amounting almost to paralysis arrests the reason, precipitating all the higher faculties into a mere animalism ;

sensual, selfish, sluggish, varied only with paroxysms of anger, that are senseless and brutal ; in appearance the beer-drinker may be the picture of health, but in reality he is most incapable of resisting disease. A slight injury, severe cold, or shock to the body or mind, will commonly provoke acute disease, ending fatally. Compared with inebriates, who use different forms of alcohol, he is more incurable, and more generally diseased. The constant use of beer every day gives the system no time for recuperation, but steadily lowers the vital forces ; it is our observation that beer-drinking in this country produces the very lowest forms of inebriety, closely allied to criminal insanity. The most dangerous class of tramps and ruffians in our large cities are beer-drinkers. It is asserted by competent authority that the evils of heredity are more positive in this class than from alcoholics. If these facts are well founded, the recourse to beer as a substitute for alcohol, merely increases the danger and fatality following.

In bitters we have a drink which can never become general ; but its chief danger will be in strengthening the disordered cravings, which later will develop a positive disease. Public sentiment and legislation should comprehend that all forms of alcohol are more or less dangerous, when used steadily ; and all persons who use them in this way, should come under sanitary and legislative control.

CHAPTER XIV.   •

CLIMATIC PERIOD—DIAGNOSIS OF INEBRIETY AND STUDY OF SOCIAL STATISTICS.

"There is a natural tendency in our very life to find a settled channel in which its forces may flow. So conscious are we in the earlier periods of our adult life of the existence of strong and opposing passions and tendencies, that we are apt to accept as an axiom the common expression that 'character is not formed until forty.' Physical factors are all the time at work to develop and mature what is in, and of us, and it is not until maturity is reached that a definite and uniform channel is wrought out and established. This age of maturity is not and cannot be determined according to a uniform rule, but just as the sexual instinct has its time of beginning and ending, so other elements have a similar history."

Again, "There comes a time in the course of one's life when the forces that have been engaged in structural repair and waste come as to a stand-still, and consult together as to which shall be dominant in the future. Some of the organs may be said to rest entirely after a certain age; rest by a cessation of function, and become atrophied also; some, that do not rest absolutely, undergo a modification of functional activity. It is at this slack-water period—at the middle of life—when the molecular deposits in the organic structure are different in quality than formerly, that we look for a different product."

The craving for drink is not a craving of childhood, it does not usually declare itself until the demands on the nervous system begin to be exorbitant, and its *terminal*

period comes with as much certainty as does its initial stage.
That terminal period is the climacteric period.

Many years ago Dr. Parrish called attention to this thought, and urged those who had opportunity to observe to note the period of life when the largest number of reformations or cures of inebriety were accomplished, and stated that he believed they would be found between the ages of forty and fifty. Subsequent observation has confirmed his view. *"Between these ages especially do recoveries that are spontaneous occur,* and statistics show that by far the greater number of persons first exhibit the alcoholic proclivity between the ages of fifteen and twenty-five, and though unable to verify the statement by figures, he was convinced that the allowance of twenty-five years of use will, in most cases, close the drinking career, either by *exhaustion of the desire,* or by the fatal termination of the individual life. He thought it will be found also, that when inebriates have lived beyond the middle period of life, so as to attain the three or four-score limit, the commencement of the drinking career was considerably later in life than the average period named."

Dr. Mason from a careful study of six hundred cases in this country and one hundred and fifty-two treated in England, found the greatest liability to be from 30 to 40, and after 45 or 50 the decline in the drink curse was very rapid. Dr. Kerr found that this period was from 55 to 60. Other observers agree that a time occurs in each case where the possibility of the subsidence of the drink curse is favored by such a period.

In the early stage of inebriety the diagnosis of the disease is extremely difficult. There are no physical signs upon which the examiner can rely. He must draw all the facts on which to form his opinion from the applicant himself or his friends, neither of whom are likely to conscientiously confess the truth. The applicant himself, as a general rule, is ignorant of the fact of disease, and will account

for his personal habits in some other way. Fortunately, if
he is not thoroughly posted in regard to the symptoms of
the disease, he will unwittingly betray the truth by the
revelation he will make of his habits.

First of all, it must be borne in mind that the inebriate
is a chronic deceiver, and cannot be depended upon to tell
the truth.

His testimony is not to be relied on even under oath
when there is a motive to deceive, unless there are corrobo-
rating circumstances to sustain it. Impairment of the moral
perceptions is one of the first symptoms of the disease, and
is also an obstacle to a correct diagnosis.

An inebriate never sees himself as others see him. If he
did there would be more hope in his case.

The heredity of the applicant should be carefully inquired
into. No man is a good risk if he uses alcohol in any form,
or to any extent ; if he has an inheritance of inebriety,
insanity, epilepsy, or any other form of neurasthenia, no
matter what a man's present habits of indulgence may be,
he will not be able to adhere to a temperate use for any
great length of time. It is a recognized fact that *three-fifths
of all the persons* who use alchoholic liquors between twenty
and thirty become inebriates ; and the inference that persons of such a heredity who indulge in alcoholic stimulants
will become inebriates, amounts to almost a certainty.
Careful inquiry will elicit the fact that such persons use
alcoholic drinks because they like them, because the effect,
if not the taste, is agreeable. A man may like claret because
it relieves his thirst, or he may like beer because it gives
him an appetite ; or he may indulge, in a company of friends,
for the sake of sociability or hospitality, and it may not be
considered a symptom of inebriety. But when it is evident
that the man drinks because the specific effects of alcohol
are especially agreeable to him, he must become a total
abstainer, or it will be only a question of time when he will
become a confirmed inebriate.

A positive dislike or aversion to the use of alcoholic drinks at certain intervals, and a strong desire for them at other periods, is often a characteristic of the inebriate diathesis. When an admission is made of an occasional indulgence to excess, it is very important to know why it is occasional, or what the occasion is. If it is periodic, or at somewhat regular intervals, it must be regarded as evidence of disease.

An abnormal desire for alcoholic drinks, however manifested, is an evidence of disease. If the diseased condition has once been established, it is always liable to return ; and the percentage of exceptions in favor of permanent cure is so very small that a period of abstinence for one or more years does not make the applicant a safe risk. Men who have had alcoholism, dipsomania, or delirium tremens, and have reformed to such an extent as to be total abstainers, for any considerable period, go down very rapidly if they return to their old habits.

Under such circumstances their prospect of life is very low ; not more than four or five years at most in the majority of cases.

All cases of inebriety may be classed as persons of *undeveloped, degenerate*, and *disordered* minds. The first class, the *undeveloped*, represents all grades of defective, retarded growths, sometimes seen in external conformations of head and body. Grades of ancestral defects and brain failures, seen in faulty acts and thoughts. The second class, of *degenerate* brains, are those who are retrograding both in mind and body. Such cases frequently date from illness, injuries, shock, diseases of all kinds, and diseases of old age, and of the nerve centers. The third class, the *disordered* brains, are those who from ill-health, bad conditions of living and surroundings, have developed inebriety. Such cases cannot bear the strains, drains, or any extreme circumstances, which tax the energies and vigor of the body. In all these cases, inebriety starts from unknown states and exciting causes, and these marked physical conditions are both primary and secondary. Inebriety is always disease and degeneration, and the fact often disputed is, can this disease be traced in any uniformity of symptoms or progress? Dr. Jackson has clearly pointed out, that all disease follows a regular retrograde march, which can be outlined and studied as clearly as growth and development. Inebriety is no exception to the rule.

The frequent instances where inebriates in apparent possession of good judgment, go away and drink to great excess, displaying a degree of forethought and premeditation fully characteristic of all the ordinary events of life, are often very confusing to the ordinary observer. When the drink paroxysm comes all unexpectedly upon the victim, in some unforeseen state and circumstance, and he falls, it is dimly apparent that he is suffering from some unstable or diseased brain state, which has burst out from the application of some exciting cause. But when the paroxysm is anticipated and prepared for, and all the surroundings are made subservient to this end, when every facility to procure spirits are increased, when money is secured and business arrangements are made in view of this coming paroxysm, the conclusion most commonly reached by all non-expert observers is that it is deliberate vice and wickedness. When the histories of a number of these cases are studied and compared, they are found to be well-marked cases of reasoning insanity, with drink paroxysm. These paroxysms are the acute attacks—the deliriums which expend themselves like storms that gather and burst—and are preceded by long periods of rest. A typical case is that of a banker, who is a man of excellent judgment in all business and social matters. He will prepare for a week or ten days in advance for a drink paroxysm. He is a temperance advocate, yet he will display great cunning to conceal the approach of this "spell." He will become very active in his temperance efforts. His friends realize his danger, and try by every means to help him, but find that all their efforts are turned to aid him in concealing it. He will not begin unless he can find some way to conceal his presence while the paroxysm is on him. When his friends were vigilant, he has been kept sober for a week or more, but with the first opportunity he disappears, and all their work has been thwarted by his cunning. In another case, the most deliberate reasoning and planning will mark the paroxysm. In another case, all at once he will rush away and drink in the most suicidal and insane way, only giving as an excuse some real or fancied injury or trouble. This deliberation and cunning is a symptom of mental unsoundness, and is most obviously reasoning insanity, and will be recognized when these cases are better understood. Dipsomaniac and periodical inebriety very soon became reasoning maniacs, dangerous be. cause they are unknown and misunderstood.

In a study of social statistics, *nativity, sex, age, temperament, climate, occupation, custom, and social conditions,* are important factors in the etiology of alcoholic inebriety, outside of the well-known causes preceding or accompanying disease or injury and heredity, and also, to a certain extent, are to be taken into consideration in the treatment of all cases of inebriety.

Nativity of the 4,663 cases under care at Fort Hamilton, N. Y., was as follows : United States, 3,186 ; Ireland, 826 ; England, 203 ; Scotland, 77 ; British Possessions, 73 ; Germany, 109 ; other nationalities, 44 ; not recorded, 145. The United States naturally compose a large majority, as we find in asylums of other countries inhabitants of those countries as inmates will be in excess of all others ; but we must acknowledge that the nervo-sanguine temperament of the American is peculiarly susceptible to the evil effects of alcohol, and that, other things being equal, the average American would sooner succumb to inebriety than his transatlantic brother. Ireland takes the lead among foreign nationalities. Next in order we have England,

Germany, and Scotland ; the lesser nationalities occur in insignificant proportions, and do not call for special comment.

As a matter of *racial importance* I cannot recall a single
instance of acute or chronic alcoholic mania in the negro
among the several thousand inebriates who have passed
under my notice during a period of nearly twenty-four
years.

In reply to a series of questions, Dr. Edwards, of Richmond, informs me " that the negro is rarely the subject of
chronic mental or nervous disease arising from alcohol,
although it is rare to find a negro, male or female, who
does not drink. Alcoholic liquors are preferred, and yet
the laboring negro, as a rule, is not a drunkard." He
attributes this to the out-door life, simple habits, and low
grade of nervous organization of the negro.

Climate.—As to the influence of climate on inebriety,
we have not any special statistics to show, but it is a popular impression that the inhabitants of low levels, especially
near the sea-coast, are more apt to be intemperate than the
dwellers on the higher plateaus, table-lands, and mountainous districts. The influence of certain barometric conditions, dependent upon atmospheric changes, in influencing
and producing certain conditions of the nervous system, is
familiar to all who have made this subject a special study.
It is also a well-known fact that in malarious districts the
depressing effects of the malarial cachexia are counteracted
by the free, habitual use of quinine, strychnine, coffee, and
other nerve stimulants, among which alcohol predominates.
No one will dispute the fact that an unhealthy, enervating
climate is more apt to produce intemperance and consequent inebriety, than a climate having just the opposite
characteristics. Climate, undoubtedly, is one of the factors
in the production of inebriety, as it is of other diseases.

Sex.—There were 4,084 males and 579 females. The
Fort Hamilton Asylum is intended more especially for
males, hence the small proportion of females. It has no

provision for females of the better class ; the female inmates were from the middle and lower classes of society, but the inference must not be drawn that inebriety does not prevail among females of all classes, or that the above is a fair relative proportion of the inebriates of both sexes. In this country, undoubtedly, the male inebriate far exceeds in numbers female inebriates. But this is not universally the case. In England and Wales the habitual inebriate females already convicted to the habitual inebriate males already convicted are as three to one. In England, especially among the higher classes of society, inebriety prevails to a greater extent among females than in the same class and sex in the United States.

Social Conditions.—The married male inebriate exceeds in numbers the unmarried male inebriate. There were 2,098 married, 1,744 single, male inebriates ; and especially does the married female inebriate in far greater proportion exceed the unmarried female inebriate. There were 401 married female inebriates and only 48 single female inebriates. Must we conclude that, other things being equal, the married life predisposes to inebriety ? It would seem so, in the case of females at least. The married female has a much greater strain upon both her mental and physical constitution than the unmarried. A fact substantiated in the reports of our Asylum further shows, that unmarried males are more frequently subjects for re-admission than the married ; that is, that they are more apt to relapse. The reverse is true in regard to females. Married females are more apt to relapse than unmarried females. The latter are not apt to relapse ; if they do, it is the exception to the rule. The spinster then, is the most temperate member of society. There were 242 widowers and 130 widows also recorded.

Approximation of Ages.—The ages of the majority of the cases treated were from 20 to 60, the greater proportion from 30 to 50, and of these considerably over one-half were between 30 and 40 years of age. Below the age of 20 and

above the age of 60, comparatively few. The oldest patient
was 73 years of age, and the youngest 18 years. We must
conclude that the great majority are of that period of life
which is the most effective for usefulness and attainment
under normal conditions. In other words, alcohol cripples
and handicaps the majority of inebriates at the most useful
period of life. Another fact is, that inebriates may exceptionally, but do *not* as a rule, attain to a *long life*. About
one in 385 of inebriates, whom we treated, reached the age
of 70 years.

Occupation.—Let us now consider the relation of inebriety to occupation. Imagine a community of 4,663 adult
inebriates, embracing every trade, employment or profession. Excluding 234 males who had no occupation and
562 females, 275 of whom were unemployed, and the
balance either domestics or housekeepers, we have about
3,867 males who were variously occupied, representing
two hundred trades, professions, commercial, mercantile, or
agricultural occupations.

We find directly engaged in the *liquor business* 71 bar-
tenders and 51 liquor dealers. The indoor trades exceeded
the outdoor trades. Among those engaged in the outdoor
occupations, inebriety seemed to affect most those whose
business especially exposed them to irregular hours and
inclement weather,—teamsters, cartmen, coachmen, carmen,
conductors, drivers. One hundred such are recorded. The
next in frequency were butchers (45) ; next stone-cutters
(28) ; next plasterers (26) ; next coopers (19). The other
occupations being at or below the latter figure, and running at or about the same average or percentage. Among
in-door occupations we find painters (121) in the majority.
The painter handles alcohol, turpentine, etc. ; his occupation is not a healthful one ; he is apt to contract diseases
incident to it, as turpentine poisoning, lead colic, and nervous diseases arising from lead poisoning, as wrist drop, or
paralysis of the extensor muscles, etc.

Next in order come printers (58) and pressmen (58). Long hours, extra work, night work, and an employment that demands great rapidity, and probably more mental and physical strain than the average occupation, may account for the fact that the printer and the pressman take the second place.

The other trades are in the following order : Shoemakers (45), plumbers (39), tailors (38), hatters (34), tinsmiths (31), waiters (28), photographers (27), carpet-weavers (22), glass-blowers (21)—the remainder of the in-door trades were at or below 20.

The greater part of the various occupations were from mercantile or commercial life. At least 1,200 or 1,300, or about one-third of the entire number belonging to one or the other of the above classes in the following order : Clerks (565), merchants (283), bookkeepers (100), salesmen (152), agents (78), brokers (44), the balance being made up of canvassers, contractors, railroad officials, bankers, publishers and superintendents, etc.

In agricultural occupations we note, farmers (34), and gardeners (15).

Professions.—Four hundred and seventy-seven, or about one-eighth of the whole number, belonged to the professions, as follows : Physicians (115), lawyers (111), engineers (58), druggists (43), journalists (39), artists (32), students (21), reporters (19), clergymen (10), actors (9), the balance being in small proportion, architects, accountants, actuaries, notaries public, chemists, assayers, army officers, dentists, editors, etc.

Why physicians are in excess of other professions is due to the fact that they lead very arduous lives, both physically and mentally, with irregularity as to sleep and diet, rest and recreation ; but there is another fact also, the physician would be more likely to appreciate and avail himself of asylum privileges for the treatment of his inebriety than any of the professions.

In considering the various avocations the usual average relation of the occupation or profession to a normal condition of society must be considered. In this way only can we get at the fact as to whether any one calling exceeds the other in a tendency to lead to inebriety.

Results of Treatment, etc.—Total cases, 4,663 ; still under treatment, 1,283 ; total cases discharged and to be accounted for, 3,380 ; doing well, 1,465, 43 per cent. ; lost sight of, 662, 19½ per cent. ; unimproved, 555, 16½ per cent. ; re-admitted, 556, 16½ per cent. ; died, 81, 2½ per cent. ; transferred to other institutions, 61, 2 per cent.

Doing well.—By this we mean the patient is restored to society, to his business and social relations. Exclusive of *death* and *transfer*, the percentage will be nearer 45 per cent. It must also be considered that the inebriate on an average is not brought to us for treatment until after *his inebriety has existed ten years,* his system broken down, and oftentime the subject of incurable disease, the result of his inebriety or some disease or injury with which his inebriety is complicated, and which may have preceded and been the cause of it. Again, not only do inebriates apply at a late period for treatment, but only about one-quarter of those who do apply remain over six months, while three-fourths remain at periods varying from one to four months, and the large majority less than the legal limit of three months, so that if we were to apply the same rule that is applied to other diseases which are submitted to us to be cured, that is, having the patient brought to us at a *reasonable period after* the disease tendency has manifested itself, and having also the patient *remain under* our care a *reasonable length* of time for treatment as each case may demand, it would be easy to see that our ratio of cures would be 75 or 80 per cent., instead of 43 per cent. as they now are, which is nevertheless a good showing considering the disadvantages we have had to contend with. We are confident that in the future, under proper conditions, at least three-fourths of the inebriates treated in our asylums will be cured.

Thirty-six per cent. were *lost sight* of, or *unimproved*, but this does not mean they will not be heard of again ; a certain proportion will return to our institution, and of these a number will be cured. Some of our most successful cases are those which have been in the asylum at different periods under treatment. Of the balance, some will die, others will move away, others go to similar institutions elsewhere located. We shall not make any comments as to *deaths* except to call attention to the *remarkably* small percentage, which is about equally divided between those who died outside of the asylum and those who died in the asylum. Some 61 were transferred to other insitutions—30 to the *lunatic asylum.* We observe the tendency of inebriety toward insanity. The great majority of inebriates carry unevenly-balanced minds ; they are on the verge of insanity all the time, and not unfrequently pass over the line. In any inebriate asylum it would be safe to assert the large majority of inebriates, at least for the first few weeks after their entrance into the asylum, are in a mental condition that, to say the least, is not normal. The above 30 transfers were marked cases of lunacy, acute or chronic mania, which were not suitable for an asylum of our character. The moral effect and the law of association forbid that the inebriate of weakened mind and body should be associated with insane persons. The tendency of every inebriate is that way, and such association would only precipitate the event. This is the principal argument against the incarceration of insane persons and inebriates in the same institution, although there are other arguments equally effective. A certain proportion of persons were brought to the asylum suffering from various diseases or infirmities that rendered them unfit subiects for our asylum. Of these, some 31 were transferred to hospitals or other institutions.

We have thus given a few general observations, resulting from a study of these special statistics. We have not by any means exhausted the subject, but we trust we have added some points of interest and importance, not only to the etiology of inebriety, but also some indications for its more successful treatment.

CHAPTER XV.

The existence of certain specially dangerous and hyper-acute forms of delirium tremens has been known and recognized by many writers since the days of Magnus Huss, but the first distinct description of this class of cases in which an attempt at their differentiation from the ordinary type was made, seems to have been that of Delasiauve, of his so-called superacute form in 1852. In his cases, however, the distinctive symptom of fever was absent, and it was left for Magnan, in 1873 and 1874, to describe as a distinct variety, his cases of febrile delirium tremens.

Although the cases which I am about to report do not agree in all respects with those related by him, I have ventured to make use of the title of febrile delirium tremens for them also. One of the predominant symptoms, perhaps we may even say their most striking characteristic, was the presence of fever, and in many points they closely resemble the cases of Magnan. Moreover, there can be little doubt that the different varieties of delirium tremens shade imperceptibly into each other and that they are all acute manifestations of chronic alcoholism modified in their symptoms by the constitution of the patient, his condition at the time, the existing complications and various other attending circumstances.

The presence, however, of so important a symptom as fever, affords, in our opinion, a sufficient ground for placing those cases in which it exists to any marked degree, in a separate category from the ordinary simple febrile cases.

Since, after as thorough an examination as possible of the voluminous literature of delirium tremens, we have found no carefully reported cases of this character, since those of Magnan, we feel justified in bringing these to the notice of the profession more especially as we believe that there are certain considerations connected with them of considerable importance both to the specialist and perhaps even more to the general practitioner.

A full description of the case is given, of which the following is a summary :

In this case we have a man of thirty-two, with a distinct predisposition to mental disease, and addicted for a considerable period to the excessive use of alcohol. Having continued the abuse of the stimulant after, at least, two attacks of delirium tremens, he finally, four weeks before entering the hospital, and probably much earlier, is perceived to act in a strange manner, to have temporary lapses of memory, and hallucinations of sight and hearing. On entrance, he is found to be much in the condition of a patient recovering from a severe attack of delirium tremens—weak, with decided tremor of face and hands, and mentally affected, unable to realize his surroundings. Very shortly, *fever* was detected, and, instead of improving, he grew rapidly worse. The general weakness increased, and was accompanied by very marked and constant subsultus tendinum, and by continual plucking at the bed-clothes. The tongue was cracked, dry, and parched, and the general condition suggested that of typhoid fever. Mentally, he likewise became worse, having constant, rapidly changing hallucinations, many of them terrifying and horrible, but many also, and the proportion of these increased as the disease progressed, of a not unpleasant character. He constantly imagined that he saw friends and acquaintances, who spoke and chatted with him. He talked aloud much of the time, often starting up suddenly and answering some subjective question. This condition of things continued for a month, being varied by periods of semi-coma, when he could with difficulty be roused to answer questions. The fever ceased at one time, to recur later. At the end of a month his strength had improved, and there were no acute symptoms remaining. From this time his mental condition improved until he left the hospital, only to grow worse afterwards, and to necessitate a still longer treatment before final recovery.

In this case, there is little doubt that there existed in the beginning that condition of chronic alcoholic poisoning so carefully described by Lentz, under the title of chronic hallucinatory alcoholism. This, possibly due to the withdrawal of stimulants, passed soon into an acute febrile condition, recurrent, and lasting about a month, to be followed in its turn by the ordinary hallucinatory alcoholic insanity.

The second case was that of a liquor-dealer, thirty-two years of age, a native and resident of Boston.

One of the patient's uncles was insane for a time, and at the Worcester Asylum for six months. No record of any other mental or nervous affection in the family obtained.

Patient had been healthy, except as follows : He has been a constant drinker for two or three years, and at times has drunk very heavily. Last autumn, he had several epileptic fits, and he has fallen down stairs several times. Two weeks before entrance, he stopped the use of alcohol entirely for five days, but then resumed. He was at this time placed under the care of a physician, as he had begun to have hallucinations of sight, and was rapidly growing worse. Finally, being inclined to be violent, he was brought to the hospital.

This case was similar to the others.

In both the preceding cases we have to deal with patients in whom, we may presume, a certain tendency to mental disease exists. One had already himself been in an asylum; the other had had a near relative in one. In both, the existing affection was undoubtedly induced by addiction to the excessive use of alcohol. The more prominent and noteworthy symptoms in these cases may be resumed as follows : (1) The duration of the disease, followed in both cases by recovery, in connection with (2) the peculiar temperature, rising at times to 102° or more ; (3) the great weakness of the patient, especially in the earlier stages ; (4) the typhoidal appearance ; (5) the constant subsultus tendinum, and the plucking at the bed-clothes ; and (6) which is less uncommon, the long duration of the peculiar form of delirium which in the beginning, precisely resembled that of delirium tremens, but continued for weeks, with a gradual change to less terrifying hallucinations.

I will not enter here into the question of the differentiation of these cases from typhoid fever and other diseases, but will merely say that in the cases related there did not exist any of the more diagnostic symptoms generally seen in typhoid fever, except the general typhoid-like condition of the patient, and that in both cases the course of the temperature was unlike that usual in typhoid.

The medico-legal importance of such cases as these seems to me considerable. The question to decide is whether the patients were actually insane, and likely to remain so for a considerable period of time, or whether the condition was a more or less temporary one, and the patient had a fair chance of recovery. In the first case, the actual question to be decided was whether the patient should at once be committed to an asylum, or whether it were advisable to wait ; in the second, whether the patient was likely to remain in his actual condition so long that it was right and advisable that his property should be put in trust. In both cases the decision was in the negative, and rightly so.

I cannot help feeling, for these reasons, that it is very important that this class of cases should be early recognized, not only by specialists, many of whom have, undoubtedly, opportunities for observing them more or less often, but also by the general practitioner, into whose hands they, in the beginning, almost always fall. All the cases of this character that I have seen, or have been able to find accounts of up to the present time, have either been fatal within a few days, or have ended in recovery. At any rate, if death does not occur within ten days, the prognosis is more favorable than one would be led to suppose by simple consideration of the duration and character of the symptoms.

In regard to the pathology of these cases, although data sufficient to justify a decided opinion are wanting, certain facts bearing on this subject may be mentioned. Without entering into detail in regard to the pathological changes of the nervous centers and their envelopes in alcoholism, we may refer to a few general results. Fournier states that autopsies after *acute alcoholism* in man show most commonly the following lesions : "cerebral congestion, more or less intense ; meninges injected, veins and vessels of the pia mater gorged with blood, cerebral substance dotted with points, roughened (sablé), and, on section, permitting the escape of fine drops of blood ; sometimes, also effusion of serum into the meninges."

In chronic alcoholism we find two classes of lesions in these organs ; the one, which may fairly be classed as acute, though the result of chronic changes, comprising, for example, hemorrhages, and perhaps some effusions ; the other class, the subacute and chronic. Audhoui, writing in 1868, says : "There is no need, I think, of insisting on the form that the nutritive trouble affects in the nervous centers, thickening of the meninges, the production of false membranes on the cranial dura mater, adherence of the pia mater to the cerebral cortex, hypergenesis of the neuroglia, fatty degeneration of the nerve-cells, of the capillaries, etc. ; all this is perfectly well-known."

In regard to the superacute form of delirium tremens we may mention two varieties, the "forme suraigne" of Delasiauve, and the delirium tremens fébrile of Magnan.

Delasiauve's form is described by Lentz as follows :

" The forme suraigne of Delasiauve is remarkable particularly for its violence, its agitation, the intensity of the delirium and the gravity of the general condition. The nervous activity is prodigious; the patient has neither respite nor repose, no part of his body is free from movement; his face bloated, red, even violet, is contorted through the quivering of the muscles; his eyes roll in their orbits; his skin is hot and burning, is moist with a profuse and sticky sweat, which sometimes emits an alcoholic odor. The tongue may preserve its natural moistness; more often it is dry along the edges, and its surface as well as the edges are covered with fuliginous crusts. Usually the thirst is excessive, unquenchable; the respiration more or less labored; the alteration of the features indicates a profound prostration. As to the pulse, sometimes rapid and feeble, at other times it contrasts by its almost normal rhythm with the other symptoms. The mind is assailed by hallucinations whose rapid succession causes an incessant change. The words crowd each other so in the patient's mouth, that several demand utterance simultaneously and escape with difficulty in jerky, interrupted, often unintellibible sentences. In constant agitation (jactitation), the head and hands are moved abruptly in all directions whence the imaginary impressions seem to arise."

Magnan's form differs but little. Its principal distinctive feature is the rise of temperature which is apt to run high and reach 40° C. (104° F.), or even 42° C. (107.6° F.). This lasts without remission for two or three days or perhaps longer, and if not followed, as is usual, by a fatal result, gradually descends to the normal limit. This form also is marked by the constant presence of muscular movements, subsultus tendinum, jerkings and contractions of the muscles all over the body, and by the extreme muscular weakness which eventually results from this incessant activity. Lentz considers that the only symptom by which this form can be differentiated from that of Delasiauve is the possibility of prolonged remissions in which the consciousness may for a time become quite clear.

The presence of a high and continued fever during an attack of delirium tremens, is always a symptom of most serious import. It denotes either the presence of some severe and dangerous complication, as pneumonia or meningitis, or it implies, as is thought to be the case at times, by certain authorities, an affection of the cerebral heat-centers, and thereby a wide-spread and dangerous condition of the cerebrum. In ordinary cases of delirium tremens, there is no rise of temperature whatsoever.

Näcke says : "In a series of examinations of a small number of cases (eleven), a slight feverishness could be determined in one-third of them. The maximum was 38.8° C., 101° F. Any temperature above this pointed to some internal inflammation, more especially pneumonia. In our cases a slight fever appeared in the evening only, as a slight rise of the physiological evening exacerbation of the temperature, never in the prodromal stage, commonly only on the first, rarely on the second day of the true delirium. Pulse and respiration were commonly only slightly increased in rate."

The cause of the fever in febrile delirium tremens is still doubtful. Magnan gives the results of five autopsies in which little definite was found beyond the injection and œdema of the cerebral meninges and a similar condition of the meninges of the spinal cord with injection of the gray substance of the latter. He himself says, that besides the hyperæmia, which sometimes ends in hemorrhage and thus attests the very violent irritation of the nervous centers, we scarcely find at the autopsy anything except the more or less advanced alterations of chronic alcoholism.

That delirium tremens may be complicated by meningitis, is, of course, well-known, and many instances have been published, of which, however, I will only refer to the cases of Bonnemaison.

Whether such complication exists in any special case, can, of course, only be decided after a careful consideration of all the symptoms.

Whether in the febrile cases ending in recovery, the fever is due to complications meningitic or otherwise, or whether it is simply due to the violence of the cerebral irritation and the affection of the cerebral heat-centers has not yet been proved. The evidence in favor of the latter condition is up to the present time wholly negative. Considering the existence of heat-centers, as shown by Dr. Ott and others, proved, since no other cause of the high temperature is apparent, and since cerebral irritation evidently exists, it is assumed that the fever is due to the irritation of these centers. It must be remembered, however, that this is only a theory with some plausibility in its favor.

Näcke is in favor of this view. As in Magnan's cases the temperature cannot be simply dependent on increased muscular action "since now at the autopsy of such patients, beyond the more or less marked hyperæmia of the central nervous apparatus, and the changes produced in the system by chronic alcoholism, nothing was found which could explain the violent fever, we must in these cases regard the fever as directly dependent on the action of the lately introduced masses of alcohol upon the heat regulators."

We, however, do not believe that this question can yet be decided without further evidence.

Another form of delirium tremens has been noticed as coming from traumatism.

This nervous affection, characterized by muscular tremor and a peculiar restless delirium, not infrequently follows the receipt of injuries in those accustomed to alcoholic stimulation.

Some writers describe, under the terms traumatic delirium and nervous delirium, a condition frequently very similar to delirium tremens, which is said to occur in patients free from the alcohol habit, and to depend upon nervous prostration, often associated with shock and hemorrhage. It is possible that failure to investigate previous habits with judicial acumen has allowed to arise a confusion between delirium dependent simply upon traumatism and delirium induced by traumatism in alcohol drinkers. The muttering delirium and muscular twitching that supervene in asthenia, from surgical as from medical causes, and the noisy delirium after injury that is exhibited by quick, rapid, and full pulse, and by febrile reaction, are two very different conditions to which the name traumatic delirium might with propriety be applied. These forms of mental disturbance, in my opinion, better called asthenic and inflammatory delirium respectively, arise without reference to personal habits.

The group of symptoms which I propose describing as traumatic delirium tremens, however, is found especially, if not exclusively, indeed, in those whose nervous systems have undergone, prior to injury, the deterioration due to absorption of alcohol. I have not been convinced by my experience, nor by my reading, that such a concatenation of symptoms can occur after traumatism in the absolutely abstemious. The amount of drinking requisite to induce the predisposition varies with the individual. The repeated ingestion of quite small quantities of alcohol may give rise to the delirious susceptibility. It is possible that a similar deterioration of constitution, and a consequent liability to trembling delirium, may be caused by the opium, chloral, and tobacco habits ; but it is difficult to differentiate these because of their frequent association with alcoholic excess.

Traumatic delirium tremens may follow even slight injuries, but compound fractures and burns seem to have a special tendency to develop this serious complication. Its occurrence should not be ascribed to the restraint imposed upon the patient's habits by the injury, but to a traumatic disturbance of a previously unstable nervous equilibrium.

Medical authorities vary in their appreciation of the causative influence exerted by sudden deprivation of accustomed stimulants in exciting attacks of ordinary delirium tremens. It is probable, however, that in a vast majority of such cases the directly exciting causes are the deficient assimilation of food, the anxiety, and the nervous strain which go hand in hand with a period of debauch, and which persist after the ingestion of alcohol is stopped. Neither is the occurrence of the malady to be imputed to the directly poisonous effect of a large amount of consumed alcohol, since acute alcohol poisoning, in persons unaccustomed to the use of alcohol, gives rise to stupor and death, but not to delirium.

Traumatic delirium tremens occurs because chronic changes in the nervous tissue or blood, or perhaps in both, have rendered the alcohol drinker susceptible to such an outbreak upon the application of any disturbing influence. The receipt of injury is a sufficient perturbing force, especially if the patient be on the verge of an idiopathic attack. It has been thought that the use of beverages containing amylic alcohol (fusel oil) especially predisposes to delirium tremens.

The alteration in nerve structure or blood, which is the essential pathological factor of delirium tremens, is unknown to us. At autopsies, an abnormal amount of serum is usually found in the substance, and within the ventricles of the brain ; meningeal congestion and hemorrhage are often seen ; the cells of the gray matter, the cerebral connective tissue, the lymph spaces and the vessels may show sclerotic or fatty changes ; and the liver, kidneys, and digestive tract may exhibit the characteristic lesions found in chronic alcoholism ; but there is nothing to which we can point as the distinctive lesion of delirium tremens.

The initiatory symptoms of traumatic delirium tremens are sleeplessness at night, and slight tremor, which is readily noticed by ordering the patient to hold out the hand with widely-separated fingers. Subsequently, restlessness, insomnia, and tremor increase, and delirium is shown.

The delirium, which is often first exhibited at night, is peculiar. The patient sees numerous small animals or insects creeping over the bed and about his person, or is pursued by some hideous spectre. Hence, he is constantly endeavoring to eject the vermin from his clothing, or trying to escape the persecutions of his tormentor. I have now under my care a patient with traumatic delirium tremens, after an open fracture of the leg, who imagines that elephants are moving over his bed and tramping on his legs. He may, in his efforts to get rid of these disgusting and distressing annoyances, leave his bed and fall from a window or down a flight of steps. The mental condition is one of depression, trepidation, and great activity. He is exceedingly restless, and is constantly chattering in a low tone, but, though he may cry out because of fear, he shows little or no maniacal excitement. He is good-natured, not prone to violence, and can often be aroused, by emphatically spoken words, to an understanding of his surroundings: but he soon relapses into the previous incessant chattering and motion. Very often a single idea recurs again and again to his delirious fancy, and not infrequently the delirium has a comical or tragedo-comical aspect. The muscular tremor is not like the twitching of tendons seen in asthenic conditions, but resembles the shakiness, from want of coördination, seen in cerebro-spinal sclerosis. Often there is hurry in movement, and the limbs or tongue will then be thrust forward with a jerk. The tremor of delirium tremens reminds me much of the movements that would be expected in an association of chorea with sclerosis of the nervous centres.

During these symptoms, the patient is unable to sleep, is incessantly in motion, and has a bright eye with dilated pupils, and an unsteady, restless look. He exhibits a moist, flabby, tremulous tongue with a whitish fur, desires no food, has constipated bowels, and passes a scanty, high-colored urine. In idiopathic delirium tremens of moderate severity there is no great acceleration of the pulse, and the temperature does not rise much above 100°, except during active muscular exertion. In those graver cases, which Magnan calls febrile delirium tremens, the bodily heat is apt to remain in the neighborhood of 102°-125°, though there is no inter-current affection to keep up the temperature, and the pulse rate is also increased. In traumatic delirium tremens the constitutional disturbance, due to the wound, affects the pulse and temperature. The patient will often remove the dressings from his wound, or subject the injured limb to violent motion without appearing to experience pain.

Traumatic delirium tremens arises, as a rule, within two or three days after the receipt of injury, and lasts usually not more than five or six days. The illusions are apt to continue during the night, even after the patient has become convalescent and quite rational in the daytime.

The peculiarity of the tremor and delirium renders the diagnosis easy. If my view of its causation be correct, the existence of the characteristic symptoms is evidence of previous habits of stimulation ; but it is not always well to mention this suspicion, nor to call the disease delirium tremens, since the patient's friends may be unaware of the existence of such habits.

Death may occur from exhaustion, coma, or some inter-current affection, and is sometimes inexplicably sudden. The character of the traumatism may determine the mode of death. Pneumonia is frequently associated with idiopathic delirium tremens. It is often, in fact, the exciting cause of the delirious outbreak, and, of course, in traumatic cases greatly diminishes the chances of recovery. When the temperature shows a tendency to remain high without a sufficient traumatic cause, and especially when the tremor affects all the muscles of the trunk, as well as those of the head and extremities, and is not arrested during sleep, the prognosis is bad. A history of previous attacks of the disease renders the outlook more grave.

In considering treatment, it is important to bear in mind that delirium tremens is an asthenic condition. There is action, but it is the activity of weakness, not of power. Depressants are therefore injurious. Five or ten grains of calomel, or one or two seidlitz powders, may be administered in the beginning of the disease, or when its occurrence is feared, because of the anorexia and gastric derangement. Concentrated liquid food, bitter tonics, and capsicum add to the patient's strength, and tend to give tone to the impaired digestive organs. Bathing and mild diuretics may be employed, in the endeavor to eliminate the alcohol that has entered the system. Chloral hydrate (gr. x–xx) with potassium bromide (gr. xxx–xl) should be given every two or three hours, as soon as sleeplessness and slight tremor are noticeable ; no visitors should be allowed in the room. If the development of the attack is not prevented, the same treatment is continued, but the doses may be increased. The object is to quiet the nervous system and induce sleep. In this endeavor an occasional dose of morphia (gr. 1-4 to 1-3,) may be combined with the chloral and potassium bromide. The excessive use of opiates is undesirable, for it is not narcotism that is desired, but sleep; cerebral congestion is induced by overdosing with morphia.

If fatty heart exists, opiates should be pushed, perhaps rather than the chloral and potassium bromide. The combination treatment with the three hypnotics allows the surgeon to diminish or increase each element according to indications. Tincture of digitalis (m. x–xxx) every two or three hours, is valuable in cases of weak but not fatty heart, where there is pallor and cyanosis with probable anæmia of the brain. Strychnia also has been recommended in delirium tremens. Mechanical restraint, with the straps and the straight jacket, is only to be adopted when efficient watching and soothing by attendants is impracticable. All such apparatus excites the patient and is very liable to interfere with respiration. The best appliance is a loose but strong garment, consisting of trowsers and shirt, in one piece, with loops attached for fastening the patient in bed. Fractures should be dressed with plaster of Paris bandages, because ordinary splints will probably be displaced by the patient. If failure of vital powers is to be feared, alcoholic stimulants in small amounts, administered only when food is given, are judicious, because in chronic drinkers digestion will sometimes not go on sufficiently without the aid of alcohol. Such failure of assimilation in delirium tremens may turn the scale against the patient. Whiskey or brandy (F. i℥–F. iv℥ during the twenty-four hours) in the form of milk punch or egg-nog, is probably the best form of administration. Many patients will not require any stimulants. Vomiting occurring in delirium tremens, is to be treated by milk and lime-water, cracked ice, effervescing drinks, bismuth sub-nitrate, pepsin and carbolic acid mixtures.

CHAPTER XVI.

GENERAL FACTS OF HEREDITY AND PREDISPOSING CON-
DITIONS.

Alcoholic Heredity, or the transmission of a special tendency to use spirits, or any narcotic, to excess, is much
more common than is supposed. In the study of a large
number of cases, several distinct groups will be apparent.

First will appear the direct heredities. Those inebriates whose parents and grandparents used spirits to excess.
The line of the inheritance will be from father to daughter,
and mother to son ; that is, if the father is a drinking man,
the daughter will inherit his disease more frequently than
the son. While the daughter may not, from absence of
some special exciting causes, be an inebriate, her sons will
in a large proportion of cases fall from the most insignificant
exciting causes. About one in every three cases can be
traced to inebriate ancestors. Quite a large proportion of
these parents are moderate or only occasional excessive
users of spirits. If the father is a moderate drinker, and
the mother a nervous, consumptive woman, or one with a
weak, nervous organization, inebriety very often follows in
the children. If both parents use wine or beer on the table
continuously, temperate, sober children will be the exception to the rule. If the mother uses various forms of alcoholic drinks, as medicines, or narcotic drugs for real or
imaginary purposes, the inebriety of the children is very
common. Many cases have been noted of mothers using
wine, beer, or some form of alcoholic drinks, for lung

145

trouble or other affections, and the children born during this period have been inebriates, while others born before and after this drink period have been temperate.

The second group of these *alcoholic heredities* are called the indirect. They are cases where the inebriety of some ancestor has left a stream of diseases, such as minor forms of insanity, consumption, and various nerve defects, which may have run through one or two generations, then suddenly develop into inebriety, with or without any special exciting cause. In such cases the moderate or excessive drinking parents will be followed by nervous feeble-minded, consumptive, or very precocious children, or eccentric and odd people who are born extremists in every relation of life. They are persons who die early, and leave a large progeny, who suffer from nerve and nutrient troubles, and neuralgia, and find in alcohol and opium a most seductive relief from all their troubles. About one-fourth of all cases of inebriety are examples of this form of indirect heredity.

A *third group of heredities* in these *cases of inebriety, are the complex borderland cases.* They are persons whose ancestors have been insane, epileptic, consumptives, criminals, paupers, and had other forms of degeneration. Victims driven along by a tide of degenerate heredity, which burst out in varied forms and phases of diseases. This class are seen among the very wealthy and the very poor. Fully one-fourth of all inebriates are of this class, and their inebriety is only another stage of profound degeneration in the march to dissolution. In these cases there seems to be in certain families a regular cycle of degenerative diseases. Thus in one generation great eccentricity, genius, and a high order of emotional development. Extreme religious zeal, or unreasonable skepticism, pioneers or martyrs for an idea, and extremists in all matters. In the next generation, insanity, inebriety, feeble-minded, or idiots. In the third generation, paupers, criminals, tramps, epileptics, idiots, insanity, consumption, and inebriety. In the fourth generation, they die out, or may swing back to great genius, pioneers, and heroes, or leaders of extreme movements.

In the study of a large number of cases of inebriates, a *physical* and *mental heredity* will appear. Thus the children of inebriates for one or two generations will be found to have, as a rule, physical defects and deformities. Bad shaped heads and bodies, an inharmonious development, retarded or excessive growths, club feet, cleft palate, defective eyesight, great grossness of organization, or extreme frailty of development. This can be seen in the observation of almost anyone, and indicates the defective nutrition and cell growth caused by injuries from alcohol transmitted to the children. The *mental heredity* from inebriate parents is equally clear and apparent to any close observation. Mental instability, and mental feebleness are common signs. Impulsive, excitable, emotional persons, who are on the two extremes, either buoyed with great faith and hope, or depressed to the verge of despair. Extravagant self-esteem, boundless faith in the most absurd schemes of politics, religion, and science.

Men and women who are called "border-liners," meaning those whose good judgment and reason alternate back and forth over the line where sanity and insanity join. They are found in the great army of the irregulars, the intellectual and moral quacks, the badly-balanced, and weak, unstable mentality. Genius and precocity often appear in these persons. They frequently come into prominence like blazing comets, dazzling for a time, then disappear in some cloud of insanity, inebriety, or other disease. This mental heredity will be often seen in the perverted nutrient tastes of children, the impulsive appetites, and dominant animal desires. With the very wealthy and very poor, these signs of alcoholic heritage are prominent. One of the reasons is the excessive nutrient stimulation from excess of quality and quantity of food, among the wealthy, and the opposite among the poor ; also the under-work and overwork of those classes.

These are only hints and intimations along the shore of a great continent of facts, which some future explorer will reveal. It will be of interest to point out some of the results which follow from alcoholic heredity. *First*, the *longevity* is *diminished.* It is impossible for a generation with this entailment to have the same vigor to resist disease and death. Exhaustive physical and intellectual exertion is not repaired and overcome so readily, and death from slight causes are more common. Thus exposure merges into pneumonia, and other fatal conditions, more quickly than in others without this entailment.

In epidemics of fevers and other diseases these children of alcoholic parents, and inebriates themselves, die first. They die from injury, shock, strain, worry, and care. In brief, this alcoholic legacy from ancestors means a short-ened life, an early death, from varied insignificant causes and general incapacity to bear the strains and drains of the ordinary activities of life. *Second*, by a wise limita-tion of nature the race with this heritage must die out. Only by a prudent ingrafting and marriage with a healthier stock can it be continued into the future. A family with this heritage is on the road to extinction, it is switched on a side track, and is moving on a down-grade of rap-idly increasing degeneration. Nature seems to often make an effort to put on the brakes and check the speed in some remarkable fecundity.

Thus in these degenerate families you will often see a great number of children who, as a rule, exhibit many of the defects of the parents, and are short lived.

The large families of children in inebriate parents may be taken as a hint of the approach of extinction for that race. In the same way, great genius in certain directions, as for instance a poet, an orator, an inventor, or a reformer, starting far away above the levels of his ancestors and sur-

roundings, are often the last members of families far down
towards the rapids that precede the final plunge into oblivion, like the flicker of a lamp bursting into full blaze before
extinction.   *Third,* where this alcoholic heredity is retarded or accelerated by the union with different currents
of heredity, very strange compounds are the result.   Thus,
if to this alcoholic heredity are united a heritage of insanity
idiocy, or any other pronounced defective influence, all
grades of criminals, paupers, and mixed insanities follow.
While most of these defects are apparent to ordinary observation, yet there is a class of defectives springing from this
soil which may be termed moral paralytics, which will be
the subject of bitter controversy in psychological circles in
the near future.   Along this frontier line the great questions
of free will and moral responsibility must be settled.   The
injury from alcohol first numbs, then finally paralyzes the
higher brain forces, which includes all the moral elements.
This paralysis goes down into the next generation as a congenital deformity, a retarded growth, in the same way that
in some families cross-eyes, hare-lip, defective hands or
legs, are seen in every generation.   This form of heredity
produces criminals of the most dangerous type ; men and
women born without any consciousness of duty, of right
and wrong, of obligation to live a moral, consistent life.
From these mixed heredities some central brain region has
become malformed and degenerate, and the victim is without power to change or comprehend the normal relations
of mental or moral life.   Many of these persons occupy
places of wealth and influence in society, holding positions
of honor and respect, by force of surroundings and absence
of opportunity to reveal their incapacity to follow lives of
truth and justice.   .   .   .

If this subject is seen higher up, other and more startling conclusions appear.   *First,* this *heredity from alcohol is intensified and increased by the misapplication of the
educational forces of to-day.*   The highest culture of the best

colleges, applied without regard to the natural capacity of
the individual, and along unphysiological lines, most clearly unfits and destroys him.   Often this higher culture is
abnormal stimulation and growth, particularly for the
entailments of past generations.

First of all, the educational systems do not always
build up healthy brain and nerve force. *Second*, they ignore
all heredity, and influences of food, climate, surroundings, and natural capacity and the result is that abnormal impulses are intensified in certain directions, and
the power of control is diminished in a positive substratum
of exhaustion from which there is no relief. The highest
modern culture applied indiscriminately to children of
inebriates, will result in their ruin as positively as any
degree of ignorance. This is seen in the inordinate self-
esteem, feeble common sense, unstable will power, extravagant idealities, and general mental dyspepsia of many college graduates. In actual life the college graduate who
has an alcoholic heredity, and is an inebriate, is more incurable than his brother who has never had a college culture.
It has been truly said that ignorance will give more promise
of longevity, and a final triumph over this heredity than the
highest indiscriminate culture of to-day.

*Another view reveals the fact that the present legal methods to
restrain inebriety*, and the result of alcoholic heredity, produce results exactly opposite.   Thus the army of inebriates
and irregulars of this family group are held accountable as
healthy, responsible beings, and confined in most dangerous
mental and physical surroundings, actually intensifying
their defects and removing them farther from all hope of
recovery. The police courts and jails are to inebriates
literal training stations, for mustering in armies that never
desert or leave the ranks until crushed out forever. A
Chinese law, enacted a thousand years ago, and in force
to-day, contains a flash of truth. When a criminal comes
before the courts, careful inquiry is made into his ancestry.

If they are found to have any of the traits common to the prisoner he is killed and they are punished. His death ends all possibility of transmitted crime, and their punishment and recorded history puts a check on any farther propagation of the evil. Common law and public opinion are far behind the march of science in a practical knowledge of this evil and the means to correct it. Not far away in the future this terrible shadow will vanish before a larger, clearer intelligence, and all our blind efforts of to-day will be found to be but a repetition of history—the stage of empiricism, quackery, and superstition, which precedes every great advance of humanity. *From a higher point of view*, civilization and the increasing complexities and changing conditions increase this heritage. Thus every new invention which changes the direction of human activities, brings greater strain on the brain and nerve force, demanding new energies, which the alcoholic heredity victim cannot give. He is unfitted and crippled for these new conditions of life by his forefathers, left dismantled and without strength for the race, and by that great law of our being is crushed out, driven out, and crowded out in the struggle and survival of the fittest.

One great fact comes out prominently in this outline review, namely, that alcoholic heredity or a predisposition to inebriety, and many other nerve and brain degenerations, will certainly follow in the next generation from the moderate or excessive use of spirits. Parents who do not recognize this fact, practically, are committing unpardonable sins, by crippling the coming generations and switching them on the side-tracks, away from the main line of development.

Another fact appears : education and marriage should be governed by a knowledge of heredity. Education should be detemined by the family physician, and have for its object to control and antagonize all the predisposition of heredity. Marriage should be under control of law, and from

the judgment of the family physician. The time is coming
when every family will have its scientific medical advisers,
and these vital questions of heredity and practical life will
be determined from a scientific basis. Still another fact
comes up prominently. The great armies of the insane,
inebriates, criminals, and paupers are largely the doomed
victims of the sins of our forefathers. Our duty to them is
to house them, to protect them from perpetuating their
defects and injuring others. Science tells us that this
army of hereditary defectives are wards of the State, and
should be housed, quarantined, made self-supporting, and
forced into conditions of healthy living. The present indiscriminate freedom of this class is a sad reflection on the
intelligence of this century. The study of alcohol heredity
furnishes not only the strongest reasons for total abstinence
in each person, but reveals the laws and forces which govern its march in each individual, revealing a wider range of
the subject. Along this line of heredity will be found the
practical solution of many of the mysteries and remedies of
this great drink problem.

In regard to the transmissibility of inebriety from
parents to offspring through different branches and generations, and its correlation with insanity, epilepsy, and other
nervous diseases, there is far less known, in spite of all that
has been written on the subject, than is needed to be
known. It is not the disease, it is the tendency to the disease that is inherited ; and this tendency is not transmitted
to all the children, but is liable to be transmitted to some
of them, and one form of nervous disease, as inebriety, in
one or both parents, may, in some of the children or more
remote descendants, re-appear as epilepsy or insanity, or
hypochondriasis. There is a general tendency to disease
of the nervous system developed and fostered under our
modern civilization and institutions, which I call the *nervous
diathesis*, and which subdivides itself into various phases of
nervous disease, such as neuralgia, sick headache, spinal

and cerebral irritation, hysteria, and hypochondriasis, as the hand branches out into a thumb and fingers.

A philosophic study of the transmissibility of inebriety requires that these other and allied disorders be studied in connection with it, and studied in this way it would be proved that in nearly all cases of inebriety, as in nearly all cases of other and allied forms of nervous disease, there was an inheritance of the nervous diathesis on one or on both sides, and usually also an inheritance, through near and remote relatives of some one of the various forms of nervous disease that spring out of the nervous diathesis.

A similarly thorough investigation of nervous diseases of an allied character would show if not as marked, yet a very decided hereditary tendency in them all, and a constant tendency likewise to interchange, reversion, and correlation of the different manifestations of the nervous diathesis. Thus hay fever in a parent may appear as sick headache in one child, as epilepsy in another, as insanity in another; in a grandchild as inebriety, in a nephew or niece as simple nervous exhaustion.

CHAPTER XVII.

FURTHER CONSIDERATIONS OF HEREDITY.

Alcohol does not alone affect those who abuse it, but afflicts more or less grave disorders upon their children. Many books have been written on this topic, but heredity in nervous diseases resembles so closely the symptoms of alcoholic heredity, that the latter cannot always be distinguished. Another cause of obscurity is owing to the fact that alcoholic excess does not necessarily come from alcoholism, and that alcoholism can exist apart from alcoholic excess. Alcoholism by itself in the ancestors does not necessarily predicate heredity in the descendants. This requires a whole group of symptoms. Many families in which excess have happened escape alcoholism, for the reason that a sufficient number of predispositions did not exist to take on a morbid form.

Historically, alcoholic heredity was known since the days of mythology. We are told that Vulcan lame was conceived by Jupiter drunk. Diogenes, addressing a stupid child, said : "Thy father was drunk when thy mother conceived thee." Aristotle declared "that a drunken mother would produce drunken offspring." The legislation of Lycurgus promoted drunkenness in vanquished nations, that it might extinguish patriotism.

· The Carthagenian law prohibited any drink but water on the day of cohabitation with one's wife. Numerous ancient and modern authorities have repeatedly called attention to the serious and varied effects of parental drunkenness in the offspring. The first serious study was by the celebrated alienist Morel. He first defined the foundations of alcoholic heredity. His successors have done no more than develop what he had discovered.

There are admitted to be two general forms : First, the homologous hereditary, or that of similitude. Second, heredity of transformation, or eccentric heredity. In the first form the progenitor gives to the descendant his tendency to alcohol, or symptoms of his alcoholism. In the second form the alcoholized mental state of the progenitor becomes transformed into the varied nervous disorders of which the body may develop. No alienist of to-day denies this direct heredity. The examples are too frequent and striking ; but it must be remembered that heredity will appear under the most varied forms. There is but little agreement among authors about the frequency of its transmission. Dr. Dodge claims that fifty per cent. of all inebriates were hereditary. Dr. Bare believes that twenty-five per cent. are inherited. Dr. Kerr thinks over fifty per cent. Dr. Magnan claims eighty per cent. Dr. Parrish eighty per cent. Dr. Crothers eighty per cent. Dr. Day seventy per cent. Dr. Mason sixty per cent.

The mere appetite for drink transmitted from the parent to child presents so many varied aspects that it is difficult to define any exact law for its manifestation. It is not rare to see it reproduced in the child with the same characteristics as exhibited in the parents. As a general law, it may be stated that hereditary alcoholism begins early, augmenting, reaching its climax, and showing its greatest activity at manhood and at the menopause. Notwithstanding what we have said, heredity does not always reveal itself by this disordered appetite for drink.

Heredity not unfrequently shows itself by irritability, instability, and a vicious moral disposition, which seems to place the sufferer under some burden to find an excitement which will relieve him from his suffering. The laws of

nerve and psychopathic heredity govern heredity drunkenness. For this reason it may be mediate or immediate,
coming direct from parents or skipping several generations
in their descent.

Apart from the question whether strong drink produces
alcoholism or alcholism produces a desire for strong drink,
it is certain that a variety of mental and nervous disorders
engenders a desire for strong drink. Unfortunately the·
control of these cases is wholly unknown to us. This constitutes the first variety of heredity of similitude, which is
the simple transmission of the defect from parent to child.
A second variety of the same form is more discussed. This
exhibits in the descendant symptoms of chronic alcoholism
without any accompanying alcoholic excess. It is well
ascertained that the drinker does not transmit the vice of
drinking, but the disease of alcoholism. Sensibility is diminished in their descendants, and they have varied and
complex symptoms of complex nervous disorder.

The author cites two cases of the homiletic type, in
which various nervous symptoms were present, but does not
deem them marked typical cases. The heterotype form is
more easily defined. Numerous observations render it certain that epilepsy occurs under a reflex excitability produced
by alcoholic heredity. The same mother with a sober man
gives birth to healthy children, and with a drunken man to
epileptic children. Statistics prove the presence of epilepsy
in the children of drinking parents. Of eighty-three epileptics, sixty have been traced to drinking parents. Youthful
convulsions are not less infrequent than epilepsy. In the
families of sixty alcoholics, there were one hundred and
sixty-nine survivors, forty-eight of which experienced convulsions in childhood. In the case of twenty-three non-al-
coholic families, having seventy-nine survivors, only ten had
convulsions in youth.

Another form of the heterotype order is seen in the hysteria and sensitive states of women. In men, under the

name of nervosisme, it offers an infinite variety of symptoms.
A certain number of these nervous symptoms leave no
doubt of their alcoholic origin. The digestive function is
especially the seat of these disorders, prominent of which is
nervous dyspepsia, vomiting, hyperæsthesia, insomnia,
dreams, enfeebled muscular force, capricious character, and
so on.

The following clinical observations make this class more
clear : A laborer, with alcoholic father, mother, and sister,
died of lung disease. He had convulsions in youth, is neuralgic, has headaches, stomach troubles, and is weakly.
His character is unrestrained, imperious, and he has fears of
becoming insane. He drinks but little wine, because it dis-
agrees with him. B, a woman with an alcoholic father.
She has vomitings every few days, although her menses are
regular. For several years this dyspepsia is more pronounced at the menstrual period. Goes often without eating, and has cramps of the stomach. Laughs and cries, and
is very emotional at times. Has a nervous laugh, and
neuralgia and analgesia all over the body. Sensibility to
heat and cold is gone, and the sense of taste and smell
greatly diminished.

These cases give only some symptoms of the varied nerve
troubles which follow from an alcoholic origin. These
states represent a continual progression from a simple neuropathic temperament to the most pronounced nervosisme.
With women this nervosity is revealed by more or less pronounced symptoms of hysteria, unaccompanied by convulsive attacks. A third manifestation of alcoholic heredity is
impulsive madness, more rare than others. These are not
explained apart from hereditary taint. Often it is the
expression of an extended neuro-psychopathic condition.
The moral perversions and atrocious crimes often seen can
only be accounted for by the alcoholic inheritance. In
addition to this and opposed to it is another form of mania,
consisting of a delirious obsession, which completely con-

trols the idea of the person, despite all his efforts to prevent it. Its approach is known to the sufferer, but he cannot deliver himself from it. It has only recently been described, and should be considered a consequence of alcoholic ancestry.

Krafft-Ebing describes two cases of this kind, both of which had alcoholized fathers. One of these cases was a work-woman. She was hysterical from youth ; once had slight loss of consciousness ; the obsession approached gradually and consisted of melancholy and religious terror, with perplexities about dogmas. These culminated in a conviction to abstain from all food and drink. From simple mental weakness to absolute idiocy there are grades of degeneration which can be ascribed to hereditary alcoholism. These affections may be congenital and primitive, or consecutive, in the sense that the slightest cerebral derangement will result in irreparable mental weakness. Lunier thinks fifty per cent. of those who exhibit moral imperfections are traceable to alcoholic forefathers. Beyond the definite cerebral and psychical derangements which we are acquainted with, there are many moral and intellectual perversions which alcoholic ancestry fully explains. These persons are the prey of moral degeneration inherited direct from their parents. From youth they exhibit the worst instincts, are cruel, vindictive, and quarrelsome. Their delights are in witnessing suffering and tormenting others. Later they are lazy, undisciplined, and vagabonds. Usually it is impossible or very difficult to educate them, and the period of puberty is the sign for throwing off all restraint. Idleness, indecision, vagabondage, moral perversion, restlessness, drink, and venereal impulses are their principal characteristics.

These persons exhibit a wonderful precocity in developing the drink habit, resulting in alcoholism more dangerous from the fact of previous disposition to degeneration. They end by falling into the instinctive moral manias, the existence among criminals, which has been so much discussed. These subjects usually pass through many jails and prisons before ending in asylums.

Alcohol is even worse in its inflictions on the physical nature than on the moral. Children of alcoholized parents are nearly always timid, feeble, pale, bad nutrition, and lessened vitality, making them an easy prey to attacks of sickness. The muscular sytem is but little developed, and they suffer from general imperfect growth in all directions. One of the consequences of alcoholic heredity is the partial and unequal development of the brain, amounting to general or partial atrophy, unilateral, and accompanied by cranial malformations. With the microcephalus the organic development is incomplete, the cranium and superior region of the head is asymetrical, the body hemiatrophical.

Another consequence of alcoholic degeneration is hydrocephalic, and also infantile paralysis. In another order of observation it is found that alcohol heredity diminishes fecundity and birth. Lippich has demonstrated that alcoholized marriages produce two-thirds less children than among those who were temperate. There can be no doubt that alcoholism affects the generative function of both sexes. The testicles undergo degeneration in alcoholized persons. The spermatic fluid shows this in the well-marked changes it exhibits, robbing it of the vitality indispensable to conception. The alcoholic cachexia, after it has attained sufficient intensity, will produce this, although the organs may not be themselves diseased.

Many examples of women are noted who have had children by their first marriage, whose subsequent union was barren with an alcoholized husband, and also the reverse. Women may become sterile by alterations of the ovaries and matrix, and abort before maturity. From this point of view alcoholism in the mother is a more serious trouble than in the father.

CHAPTER XVIII.

Dr. Matthews gives the following statistics of heredity : " The passion for alcoholic stimulants, if not reproduced in the immediate descendants, may show itself in the successive generations, and in all cases is the most prominent factor in insanity, epilepsy, idiocy, hypochondria, hysteria, neuralgia, nervous degeneration, and its kindred ailments— often manifesting these maladies in a vicious circle—with the effect of exhibiting insanity in one, epilepsy in another, intemperance in a third, idiocy in a fourth, hypochondria in a fifth, hysteria in a sixth, and so on until the circle is completed, each generation increasing in numbers, and contributing in a direct ratio to the filling of our jails, penitentiaries, inebriate asylums, insane retreats and poorhouses. That this is not a conjectural statement the following facts will abundantly prove : In a Swedish asylum it was found that 50 per cent. of the patients had been addicted to the use of alcoholic beverages. After the removal of the heavy tax on alcoholic drinks in Norway, the percentage of increase during eleven years was : In mania, 41 per cent. ; melancholia, 69 per cent. ; dementia, 25 per cent. ; and idiocy, 150 per cent. Of the last, 60 per cent. were the children of drunken fathers and mothers. In the insane hospital at Vienna, Austria, probably one of the largest in the world, the superintendent informed me, personally, that from 50 to 60 per cent. of the insanity was due to spirituous liquors. This percentage in a country

161

where it is claimed alcoholic drinks do no harm, is well worth noticing.

In our own State insane asylum, of the now present inmates, numbering 364, 75 per cent. can be ascribed to habits of intoxication, either on their part or that of their ancestors. I am also authorized in making the statement that fully two-thirds of those persons requiring aid from city and State are descendants of inebriate parents. In one of our prominent lunatic asylums 637 cases were traced to intemperance as the assignable cause of their insanity. The statistical accounts of the State of New York give the following facts: In the poor-house of Ontario county there were 113 inmates. These, together with their ancestors for three generations, living and dead, represented 90 families, and in these families there were 168 dependents, 26 insane, 12 idiots, 103 inebriates. In Columbia county, 118 inmates, representing 114 families, had 143 dependents, 12 insane, 32 idiots, 127 inebriates. In Yates county, 32 inmates represented 26 families, of whom 59 had been dependent, 4 insane, 2 idiots, and 31 inebriates. In Kings county, 1,876 inmates represented 1,668 families, 2,039 dependent, 755 insane, ⋆3 idiots, and 973 inebriates. Herkimer county had 77 inmates, representing 67 families, 128 dependents, 21 insane, 12 idiots, and 64 inebriates. The total in the alms-houses of the State was 12,614 inmates, who represented 10,161 families, whose members for three generations, living and dead, had among them 14,901 dependents, 4,968 insane, 844 idiots, and 8,863 inebriates. In round numbers, here are 10,000 families who have produced 15,000 paupers, or 3 paupers for every 2 families—of insane, about 1 for every 2 families; of insane, inebriates, and idiots combined, about 15,000 or 3 to every 2 families."

Dr. Wright asks: Is there a form of inebriety allied to those neuroses which mark a constitutional defect in nerve balance, a defect, the conditions of which are transmissible by heredity? Some noted gentlemen deny the reality of the dipsomanical diathesis, and especially deny that such a constitutional proclivity can be handed down from ancestry. The real importance of the subject causes one to inquire with a good deal of interest : "What do the highest authorities say on these points ? The testimony of such eminent physicians as Hughes, Parrish, Crothers, Mann, Mason, and others in this country ; and Clouston, Mitchell, Kerr, Peddie, Cameron, Carpenter, and a host of others in Great Britain, any one of whom is fully competent to decide on the merits of the question, and who have conscientiously studied inebriety for some years, will not be taken. Their great interest in the subject might expose them to the imputation of prejudice in viewing the facts. Authorities who are supposed to contemplate scientific principles through an atmosphere pure and uncolored by sympathy, are the only ones who will be questioned. The attempt will be made to show that if epilepsy, spasmodic asthma, prolonged neuralgia, hysteria, suicidal melancholia and the like, are constitutional diseases, then dipsomania, or as it is called by some, inebriety, is also a constitutional disease.

Morel, when speaking of the degeneration and final extinction of a neurotic family strain, gives a history from his own personal observation thus : " *First generation—* immorality, depravity, alcoholic excess, and great moral degradation in great-grandfather who was killed in a tavern brawl. *Second generation—*hereditary drunkenness, maniacal attacks ending in general paralysis in grandfather. *Third generation—*Sobriety, delusions of persecution, and homicidal tendencies in father. *Fourth generation—*defective intelligence, mania at sixteen ; transition to idiocy ; generative functions feeble ; sisters imbecile ; wife had a bastard child of good constitution." When treating of heredity, Maudsley divides the subject into three branches : 1st, heredity of the same form ; 2d, of allied form ; 3d, with transformation of neurosis, as when the ancestral defect was

simply a nervous disease. Of heredity of the same form,
this author says : " That is, when a person suffers from the
same kind of mental derangement as a parent, which he seldom does, *except* in cases of suicide or dipsomania." Dr. M.
says in another place : " This mingling and transformation
of neurosis which is observed sometimes in the individual,
is more plainly manifest when the history of the course of
nervous disease is traced through generations." The book
of inebriety is open everywhere and to everybody. But to
study its pages aright, particularly in respect to its habits
of descent through individuals and generations, the observer
himself must abide for a long series of years amongst one
and the same community of people. It is obvious that in
this way only can he study the facts respecting the influence
of alcohol upon individuals and upon families.

In the work already cited it is written : " With respect
to an individual's legacy from his parents, he inherits not
only their family nature . . . but something from their
individual characters, as these have been modified by their
sufferings and doings, their errors and achievements, their
development or their degradation."

Dr. Blandford (*Insanity and its Treatment*, p. 139), speaks
as follows : " As I have said, the particular character of
the mania or melancholia depends on the constitution of
the individual, . . . and the same person may at one time
be maniacal and at another melancholic. It is true, we frequently see the same form in successive generations, *e. g.*,
suicidal melancholy and hereditary drunkenness. He
teaches that there is a vast number of cases where the
descending form is different from its parent ; and that the
same form may, or may not, appear in posterity. The idea
that the hereditary transmission of different but interchangeable neurotic form is an indication of the insane
temperament is universal amongst alienists. That "alcoholism is more liable to produce epilepsy, or idiocy, than to
repeat itself," places alcoholism (or inebriety) at once

amongst the hereditary and insane neuroses. "Hence, I believe springs the ever renewed insanity inebriety of our lower classes . . . My opinion is, that amongst the lower classes of our countrymen, insanity is on the increase. . . . There is a degree of drunkenness among the lower classes of this country that is not found in the higher. . . . The amount of drunkenness is enormous, and is almost confined to the lower orders—below the shopkeeper class."

This is placing habitual drunkenness, or rather the neurotic mood which craves intoxication, very closely in alliance with those neurotic states which immediately interchange with true insanity, it does indeed classify them together as one family. The intense desire for intoxication which is distinctive of the dipsomaniac, appears to issue from some unstable or abnormal state of the brain, as other lawless and unmanageable nervous symptoms frequently do, speaking of the influence of parentage in impressing a morbid diathesis upon posterity, uses this language : " The causes of insanity may come into operation at the period of conception. We should expect this *a priori*, and experience appears to prove it. We allude more especially to the case of a parent begetting children when drunk."

" When mental disease is transmitted, does the form of insanity descend ? Very frequently this appears to be the case," says Dr. Tuke. The doctor then proceeds to give examples from various authorities of the direct descent, from ancestry to posterity, of hallucinations, monomania, melancholia, mania, general paralysis, and idiocy ; and then he adds upon his own authority : " Of dipsomania, the cases are so common that it is not necessary to detail any examples." Dr. Tuke gives a table " Exhibiting the proportion of hereditary cases *in the different forms of insanity*, observed in the Crichton Institution, as reported by Dr. Stewart." Mania descends as such in 51 per cent. of cases ; melancholia, 57 per cent. ; monomania, 49 per cent. ; moral insanity, 50 per cent. ; idiocy and imbecility,

36 per cent. ; dipsomania, 63 per cent. ; general paralysis, 47 per cent. ; dementia and fatuity, 39 per cent.

Dr. Wynter says : "Among the more special forms of moral perversity, or as the alienist physician would say, insanity, which are transmitted by an insane parent, may be mentioned dipsomania." Again, the same author speaks of "the known fact that persistent drunkards plant the seeds of insanity and the other allied diseases in the offspring. Once planted there, the fruits may be diverse ; in one, there may be persistent neuralgia ; in another, the ancestral drunkenness may assume the form of dipsomania ; while another may be affiliated with partial paralysis or with epilepsy."

Dr. Forbes Winslow declares : "I maintain, and facts clearly demonstrate my position, that there is a vast amount of crime committed by persons who occupy a kind of neutral ground between positive mental derangement and mental sanity. I do not support the dangerous opinion that *all* crime is referable more or less to aberration of mind, but I do affirm that in estimating the *amount of punishment* to be awarded, it is the duty of the judge, not only to look at the act itself, but to consider the physical condition of the culprit, his education, moral advantages, prior social condition, his early training, the temptations to which he has been exposed, and, *above all*, whether he has not sprung from intemperate, insane, idiotic, and criminal parents."

Dipsomania is declared by authorities to be one of the hereditary transmissible neuroses, which is interchangeable with other neuroses that are undisputably allied to insanity. Dipsomania is an aptitude for intoxication, but it is an aptitude that is not subject to rules or laws. It is pathological for it is insatiable, unmanageable, in fact, an outlaw.

Dr. Elam asserted : "*The offspring of the confirmed drunkard will inherit either the original vice or some of its countless protean transformations.*"

Dr. Anstie affirms : "In the course of a large experience of alcoholism among hospital out-patients, I have been greatly struck with the number of drinkers who have informed me that their relatives either on the paternal or maternal side have been given to drink ; my own experience has led me to a firm conviction that particular causes of nervous degeneration affecting individuals, do very frequently lead to the transmission to the offspring of those persons of an enfeebled nervous organization which renders them peculiarly liable to the severer neurosis and which also makes them facile victims of the temptations to seek oblivion for their mental and bodily pains in narcotic indulgence. I believe that things often work in a vicious circle to this end, and that the nervous enfeeblement produced in an ancestor by great excesses in drink, is reproduced in his various descendants with the effect of producing insanity in one, epilepsy in another, neuralgia in a third, alcoholic excesses in a fourth, and so on. Among the higher classes, where it is easier than in the case of the poor to obtain tolerably complete family histories extending over two or three generations, careful inquiry elicit facts of this kind with surprising frequency. So strong is the impression left on my mind by what I have observed in this direction, that I am inclined to believe that the great majority of most inveterate and hopeless cases of alcoholic excesses among the higher classes are produced by two factors, of which *the least important* is the circumstance of external momentary temptation, in which the person has been placed where the *more momentous and mighty cause* is derived from an *inherited* nervous weakness, which renders all kinds of bodily and mental trouble specially hard to be borne.

Dr. Wright says : " Observation teaches no lesson more clearly, unmistakably and universally, than the absolute certainty of heredity. Nothing escapes it—neither the slightest physical formation or movement, nor the faintest hint of a mental or moral impulse. Race propagation is simply the handing down of the salient features of one generation to succeeding ones. There is no necessary progression, deterioration, or change. But heredity, in its highest sense, is particularly concerned and exact in transmitting an immense number of personal characteristics derived from the innumerable associations, good, bad and indifferent, that have been connected with individual existence.

"The main features of humanity are one and unvarying ; but the acquired "peculiarities" of families, and especially of individuals, are the reflections of the possibilities of human life. They measure the capabilities and the defects of man's nature, as they are susceptible of development by personal contact with the ever-changing situations, and the endless accidents of time. To the descent by heredity of the acquired characterics of individuals in varying forms, and by unnumbered vicarious substitutions, we owe the strength as well as the weakness of the race. The potentialities of humanity, exhibited in the beauty of one, the eloquence of another, the artistic genius of another, the fortitude and patriotism and unselfishness of others, illustrate and also demonstrate the existence in the human race of qualities far greater and more glorious than the highest achievements of minds merely symmetrical can possibly evince. These potentialities are the outcome of personal traits formulated in individuals by contact with the experiences of life, and transmitted in endless variety of combination and form by the power of heredity."

Dr. Kerr : " This heredity may be regarded as two-fold. There is the direct alcoholic inheritance. That drunkards beget drunkards is an axiom dating from very early times. There are also large numbers of children born with an inherited and extremely delicate susceptibility to the narcotizing action of alcoholic intoxicants, whose parents were not in the habit of getting drunk, but drank regularly and freely, physiologically intemperate, though considered by the world to be models of sobriety.

" There is also the indirect inheritance of alcohol. Under this category are ranged individuals who have no special proclivity to excess, who have no direct though latent proneness to inebriety, but who are weighted from their birth by a controlling power too feeble to stay the advances of alcohol within their very being. Alcohol, if it gain an entrance into such constitutions by however tiny an inlet, slowly yet steadily widens the aperture by increasing in volume, as the dykes built to resist the encroaches of the ocean, till all the defenses are swept away by the overpowering and overwhelming flood. In no inconsiderable proportion of cases this defective power of control is the product of alcoholic indulgence on the part of one or both parents.

" I have seen the alcoholic habit in the parent bear diverse fruit in the persons of the offspring, one sister being nervous, excitable, and inebriate, a second consumptive, a third insane ; one brother an epileptic and a periodic inebriate, the second in an asylum, the third a victim to chronic inebriety. In another instance, where all the four children have become habitual drunkards, the grandfather had also been addicted to excess.

" The heredity is sometimes crossed. The daughters of a drinking father, and the sons of a drinking mother may be the only children affected with the inebriate taint.

" The heredity may be either insane or inebriate. As inebriate parents not unseldom beget insane offspring, so from insane parentage we sometimes get inebriate children. The heredity may be of some other type. Any transmitted disease or effect of disease which increases nervous susceptibility, unduly exhausts nerve strength, and weakens control, may bear a nervine crop, in the form of asthma in one child, hysteria in another, epilepsy in another, idiocy in another, and inebriety in another."

Dr. Dacaiene concludes an exhaustive study as follows:

" 1st. Under the name hereditary alcoholism is included the totality of the pathological manifestations transmitted to a child by one or the other of his parents who are drinkers, and sometimes both.

" 2d. The inheritor of this taint, as well as the drinker himself, can hand down not only his own vice, but a special morbid tendency, a particular neuropathic state, which can always be charged to inebriety.

" 3d. The alcoholic inheritance may at first be dormant. When it exists it shows itself in infancy, or later, or in another generation. It shows itself as congenital paralysis, convulsions, epilepsy, hypochondriasis, idiocy, etc.

" 4th. The increase in the number of the insane, of the number of suicides, of crimes and misdemeanors, such are the results of hereditary alcoholism.

" 5th. It is in hereditary alcoholism that can be found the explanation of certain monsters who come from time to time to horrify society and scandalize the courts of law.

" 6th. These degenerate beings are smitten with sexual impotence. The female inebriate is apt to abort, and lastly the mortality of the newly-born among drinkers reaches a figure truly frightful.

" 7th. It has often been proved that in the case of drinkers, there is a loss in stature and physical force.

" 8th. To sum up hereditary alcoholism as well as the acquired, determines an enfeeblement of the species, the destruction of the family, and the degeneration and abasement of the race.

" 9th. From a medio-legal point of view, the hereditary inebriate, in particular the dypsomaniac, should be regarded most of the time as irresponsible, or at least his responsibility should be regarded as very limited. He is a sick man, who should be cared for, remembering that he presents an undeniable propensity to sickness, that he possesses a defective intellectual organization, in a word that he is a degenerate. If the moral sense has not completely disappeared in his case, at least its use is not accurately regulated. The judge then ought to take into account this moral state in appreciating his acts.

Dr. Lykke, Copenhagen, has shown clearly that heredity is a powerful factor in the origin of dipsomania. He indicated that psychical abnormities in the parents are developed in the next generation, and that if an improvement of the morbid condition is not attained by intermarriage, the result will be complete degeneration. One of the prominent symptoms of such degeneration is a lack of moral capacity, and the preponderance of ungovernable instincts and dipsomaniac tendencies. Hence he supposes dipsomania to be a symptom of hereditary degeneration or insanity.

Brühl Cramer, from a long examination of this subject, concludes that drunken parents are seldom prolific, and when so, the children are stupid, malicious, and full of mental defects. Skaë and Thompson have both made deep researches into the hereditary disposition. The cases which they examined were largely of the mixed form of dipsomania, and all showed decided tendencies to insanity. Skaë found in eighty-two, thirty-two cases of inherited dipsomania. In the parents of the children of the collateral branches, he found drunkenness, dipsomania, suicide, and other mental diseases. Thompson found in twenty cases, nineteen in which inebriety was clearly inherited. Many of these families containing two, four, and eight members, were all either drunken, epileptic or insane. In three families which I have studied, where both parents were drunken and insane, every single member following suffered from mental defects, of which epilepsy, dipsomania, and suicide were most common.

All authors agree that heredity is the prominent cause in the etiology of dipsomania; but they disagree as to the place it should occupy in the nosology, and whether it should be considered as a principal disease or only as a symptom. It is clear that insanity and nervous diseases develop into dipsomania in the next generation, and *viceversa*. The intimate relation between these diseases is evident.

CHAPTER XIX.

GENERAL CONSIDERATIONS OF THE PATHOLOGY OF INEBRIETY.

When a considerable portion of alcohol is consumed at one time by a person perfectly sober, certain characteristic appearances will usually be observed to follow immediately. There is a decided shock throughout the whole system. The face will often be very pale. This will be more noticeable in nervous and sensitive persons. The muscles of the face will be drawn, and fixed in position. The eyes will be bright and glittering, while their movements will be quick and constrained. The mouth will be firmly shut ; but when an effort is made to speak, the lips will be spasmodically affected. The breath will be short and panting, the pulse accelerated, and the articulation interrupted and difficult. The entire body will be affected by a trembling movement and a sensation of shivering. In fact, there will be a brief nervous chill. All these features appear at one and the same time. The period of their duration is very short, however. After the lapse of five or ten minutes at most, the second stage, characteristic of heavy drinking, will come into view. Movements will appear unnatural, and very quickly will seem absolutely distorted and staggering. Intellectual activity, also, will speedily appear both irregular and unsteady. The voice very likely will be elevated, and an incessant chatter of speech, laden with absurd boasting, will din the ear. The brim of the hat will perhaps be

thrown upward, or the hat itself placed upon the side of
the head, or thrown backward, revealing the noble forehead.
Indeed, the conditions are now all present for a full display
of aggressive vulgarity and foolishness.

The ruffianly instincts being aroused by alcohol, the
garb of the ruffian is put on. For truly it is a fact, that
when a man, being sober himself, arranges his attire like
that of a rowdy, as thrusting his pantaloons inside his boot
tops, turning upward the brim of his hat both in front and
behind, and tying a handkerchief loosely about his neck—
he will begin to feel like a rowdy, and to act like one. He
will, perchance, stick his fists down deep into his pockets,
while, possibly, he pours forth profanity and tobacco-juice
in equal volumes.

This description, of course, is not of universal application. In different persons there are considerable modifications in these particulars, owing partly to peculiarities of
individual constitutions, and partly to the kind of alcohol
contained in the liquor that has been taken. But, let the
constitutional disposition be what it may, the full drink of
alcoholic liquor will induce displays in movement, mind,
and morals, that are unworthy of the individual ; and, the
liquor being the same, these displays will, in a given individual, always be practically the same.

Differences in the kind of alcohol will doubtless occasion
considerable difference in the conduct of those who indulge
in the alcoholic habit. Dr. Norman Kerr, in his very able
work on Inebriety, says : " All alcohols are poisonous.
The least poisonous are the alcohols of wine. More poisonous are the alcohols of beet-root. Still more deadly are
the alcohols of corn (all kinds of grain) ; and the most
potent and pestiferous are the alcohols from potatoes.
Cider inebriates are usually more heavy and stupid than
alert and offensive. Amylic alcohol is nearly four times as
poisonous as ethylic." Amylic alcohol quickly brings on
muscular tremors and delirium tremens—whereas the

ethylic does not readily produce such effects, if it does at all. It is therefore perceived that while there is a sameness in the physiognomy of conduct under the influence of alcohol, there is likewise a certain difference in details, in accordance with the particular kind of alcohol that is taken. A drunken man will not be of so unruly and satanic temper after partaking of wine, or of stronger drink derived from wine, as he will after partaking of beet or potato whiskey; and he will come out of his drunken state more quickly, and with less distress in the former than in the latter contingency.

There are many persons who drink alcoholic liquors without any driving impulse to intoxication. It is a common belief that alcohol adds to the natural powers of the organism; and hence, alcohol is frequently taken in order that, through an intensified capacity, ends may be secured that would otherwise be impossible. In this respect, the occasional drunkard is to be distinguished from the spasmodic or impulsive drunkard, whose whole aim is to secure intoxication for its own sake, and also to secure it quickly and completely.

Drinking from association.—There are many persons who drink alcoholic liquors only as accidental contingencies and opportunities offer. They do not purposely seek the means of intoxication. The motives which actuate them in drinking are wholly derived from circumstances external to themselves and which are truly fortuitous. Yet the occasional drinker cannot be said to be devoid of even strong motives for his indulgence. Sometimes a man will drink alcoholic liquor simply because he happens to be in company with others who are drinking. The natural, but occasionally idle sympathy, which is so apt to bind men together in a common course of conduct, is sufficiently powerful to lead to a community of action, even in the matter of drinking liquors. Such cause for drinking may be operative on public days, or in companies that are engaged in enterprises wherein there is unity of feeling and purpose—and, of course, good fellowship—as, at log-rollings, barn-raisings, and the like.

Drinking from this cause is not apt to lead to very serious consequences. It is true, however, that if there is any considerable constitutional irritability of nerve in an individual, even an accidental indulgence may ignite a flame that can never be extinguished ; and in this way the occasional drinker may become an habitual drunkard.

The pleasurable sensations of early drunkenness are not so pronounced in the occasional drinker as they are apt to be in the spasmodic inebriate. This might be expected when it is remembered that an exquisite sense of mental and physical delight is one of the ruling inducements to frequent intoxication—that is, to intoxication for its own sake. In a stolid mind, rage and hate are not unlikely to be aroused by alcohol, instead of generosity and good temper.

Misfortunes, either domestic or in business, often lead to drinking. The benumbing influence of the alcoholic potion renders callous the distressed mind and quivering nerve. It is not always for excitement that alcohol is taken into the system. It is sometimes taken to secure repose ; and this repose is simply paralysis, more or less complete. In the paralysis of sensation, pain is abated ; in the paralysis of the co-ordinating nerve centers, moral and sympathetic afflictions no longer harass the mind. Alcohol is a complete remedy ; for the paralysis of alcohol extends throughout the whole body. It is seen in the motor system through the staggering gait, the imperfect articulations, the distorted countenance. It is perceived in the organs of sensation through the general numbness and the absence of the sense of feeling. The intellectual powers exhibit the paralyzing properties of alcohol, through confusion of mind, distortions in ideas, and in the irregular operations of the imagination. The same paralysis is seen in the moral sense through the loss of that sense, and the inflow of untruthfulness, deceit, and prevarication.

A death in the family is followed sometimes by deep potations on the part of the survivors, with the object of inducing a forgetfulness or partial unconsciousness of trouble. Yet the time thus consumed is expended in vain. It cannot serve to shorten the period which truly is requisite to assuage the intensity of sorrow. In truth, when this drunkenness has passed away, the poisoned nerves are in a pitiable state of excitement and tremor, much greater than oppressed them in the first place. With renewed distress and intensified grief, the mind is compelled to await all the longer for consolation and repose. It is in this condition of mind and nerve—after drinking is abated—when sometimes suicide is invoked to vanquish once for all the combined horrors of grief and intemperance.

Alcohol invoked to increase the power of physical endurance.— The casual drinker takes alcoholic liquors oftentimes with the view of increasing the powers of bodily endurance. The acuteness of the feelings being subdued by the nervous torpor superinduced by alcohol, the sensations of cold and heat are not keenly presented to the mind. Yet this fact does not give the physical organism any immunity from the extreme effects of changes in the temperature. Experience has shown that persons exposed to cold, wet, and fatigue, sustain themselves much better without alcohol than with it. The truth is, that alcohol reduces the temperature of the human body, as the thermometer fully demonstrates. Therefore, the drunken man perishes from cold more readily than the sober man. For, not only does alcohol abstract oxygen, the source of heat, from the blood, and thus reduce the fires of ordinary physiological combustion, but it benumbs and paralyzes certain nerve centers, whose office it is to preside over the regulation of animal heat. As a remedy in violent fevers, alcohol is extensively used for reducing the dangerous heat of the blood. As a trickster and fraud, nothing can exceed alcohol, if indeed, it can be equaled.   He who relies upon alcohol to warm him when cold is cheated and deceived.

Alcohol as sustaining muscular efficiency.—Men frequently take ardent spirits with the notion that it will increase the bodily strength.   The professional athlete, however, who understands the subject, will carefully avoid alcohol when upon the verge of action.   There are several sound reasons for this.   (a) The benumbing effect of alcohol upon the nervous powers is universal.   It includes the muscular system in common with all others.   The readiness of muscular contraction—its exact concord of action throughout the whole of its substance, and the completeness of its movement, all depend upon the natural strength, quickness, and sharpness of the nervous influence which is brought into play.   But this influence is dwarfed and minimized by the paralyzing impression of alcohol.   (b) The heart in intoxication is thrown into increased action to the extent that the additional duties which very great muscular effort would impose upon it, are too great for its capacity.   (c) The lungs refuse the breath requisite to extraordinary muscular effort when a man is drunk ; for in drunkenness they have a double duty to perform.   They are busy, not only in exhaling the alcoholic poison, but also in disposing of the deleterious material ordinarily thrown out by them. Consequently, when uncommon muscular effort throws upon them an excessive volume of blood, they are unable properly to dispose of it.   The man speedily gets out of breath, and is compelled to moderate his efforts.

The base ball player is sadly deficient when under the influence of liquor.   He cannot judge correctly with respect to distance, nor tell the true direction or velocity of a flying ball.   He can neither catch nor throw with his usual accuracy.   His eyes are wanting in alertness of action, if not in correctness of vision.   His movements and his judgment are equally at fault.   The disabling powers of the alcoholic potion are plainly displayed in the destruction of the normal relationship which should exist between the great divisions of human nature—mind and body. The perceptive faculties, as observed through the operations of common sensation and the sense of sight, are benumbed, while the muscular alacrity answering the calls of volition and judgment is absent. In brief, the athlete has not voluntary control of his muscular powers when under the dominion of alcohol, for paralysis, in whatever degree it exists, withdraws function in a corresponding degree from volition. Yet alcohol is wonderfully complicated as well as positive in its activities—volition itself being dull and paralyzed as well as the muscular system.

Similar considerations are applicable to almost all athletic or muscular exercises. The swordsman and the pugilist must possess not only a steady but a correct nerve. It is indispensable to success that the eye and the hand be untrammeled and alert, but they should also be in exact harmony with each other in quickness of perception and movement. The details should be right, and the great movements of the system, which combine the details, should be unimpeded. The prize fighter knows the fact that the universal dullness of nerve wrought upon the organism by alcohol is fatal to his efforts for victory. It is amazing to see the absolute confidence with which a person under alcoholic influence views his powers, when the knowledge of the disabilities imposed by it are not recognized. In him, all the nagging asperities of nervous activity have disappeared. Nothing seems impossible to the transcendent egoism of a drunken man. The very suggestion of unfitness or mistake is scouted by him. Equally unable to foresee, or appreciate difficulties, he has no faith in their existence.

One A——, a physician of ability, while considerably intoxicated, was handling a pistol, and accidentally discharged it. The ball flew near his wife's head, and shattered a looking-glass behind her. He was astonished exceedingly, and, with respect to this matter, he remains so to this day. His confidence in the absolute infallibility of his care and caution was simply impregnable—as it is, indeed, in every man who is drunk. Yet in this very thing he was terribly deceived and beguiled by alcohol. The sense of feeling in his hand was benumbed. He grasped the pistol with more force, and pressed upon the trigger more firmly, than he had any idea of, by reason of his impaired sense of feeling. The truth is, the confident approach of a drunken man is always amazing. His sense of feeling is dulled, and he seizes the person of another in a rude manner. His grasp is painful because it is violent. The inebriate unconsciously exerts considerable force in his movements, in order to feel that he is really in contact with things or persons exterior to him. In the case of A——, in whose hands the pistol was discharged, the harmony of action between the eye and hand was deranged, through deadening of the nervous sensibility. The consequence was that, what he supposed was a normal community of action between distant, but auxiliary parts, was an unnatural and inharmonious relationship. It was in obedience to that unrecognized condition, affecting alike mind and sensation, that the explosion took place, and not from a condition thoroughly regular and under the control of volition.

How often has a drunken man had cause to bless his "good luck," when a gun went off without dealing death, while in his hands ; and how often again, has regret and remorse followed him through life, because some weapon has been accidentally fired by his awkward and trembling movements—maiming, or possibly killing some friend or companion.

A young man, Samuel L——, recently shot a female companion through the head, killing her instantly. They were both partially intoxicated. The pistol was of cheap pattern, and was very unreliable about the trigger. The shooting was claimed to be an accident. The young man was no doubt in a state of more or less muscular incapacity. Testimony was given that in handling a pistol by a person drunk, it would more likely to be accidentally discharged than it would be in the hands of the same person when sober. Although in this case other facts pointed to murder, yet such a plea in defense might have been perfectly good and proper.

A drunken man cannot dance. It does not require a very considerable degree of intoxication to disable a person for dancing. Anything which requires a community of action amongst a number of muscles—anything which is in the line of auxiliary aid, or help in muscular actions, is an utter impossibility for the individual who is intoxicated. Yet, if there is a person in the world who thinks he can exhibit the consummation of grace, ease, and *eclat* in dancing, it is the man who is drunk.

In consequence of the inequality of alcoholic paralysis, even on the muscular system itself, the really drunken individual moves as though he was about to fall in pieces. The inebriate dancer would probably move off in sections or separate parts, were it not that his physical body is securely fastened together in spite of himself. The want of harmony between mind and body, often present, also, no doubt, disables the drunken dancer. Sometimes the muscles act tolerably well in the intoxicated individual, while his mind and will may be in a state of sad confusion. At other times his mind may be fairly clear, while he is limp and helpless on his legs.

The musician and the actor likewise fail in the exercise of their callings when intoxicated. The actor is totally unable to depict character by facial expression. The muscles of his countenance are withdrawn from voluntary control. The feature he wants to supply will not come at his bidding—it is something else, something quite different,

and yet he is not conscious of the fact.   He is deceived by
the power of alcohol, and resents the obtuseness and unfair-
ness of his critics.

The effects of alcohol upon the body of even the casual
drinker are always those of a traitor—they always betray.
The steadiness and tone of nerve which seem to follow from
its use are only indications of paralysis and insensibility
which, deceiving the mind, lure the unwary drinker into
danger and disgrace.

The late Dr. Jewell gives the following description,
which applies in many cases of inebriety : " There is a real,
pervasive nutritive lesion of nerve structure in neurasthenia.
Waste has morbidly preponderated over repair.   The nervous system becomes lean.   This involves, necessarily, loss of
power and undue sensitiveness or "shakiness" of the nervous system.   It is more easily excited than when healthy,
and the excitement when once produced radiates farther,
and endures longer, than in health.   This state or physical
condition of the nerve structures differs in degree in various
cases or at different times in the same case, from a very
slight exhaustion to that which is the most profound.   It
may extend to the whole nervous system or to only a small
part of it, or, if to all parts, unequally so to different parts.
It may be in part hereditary, or it may be acquired.   It may
affect chiefly the bodily functions, or it may affect chiefly
the mental functions, as in some cases of melancholia, or it
may involve both at the same time in varying ways and proportions, as often happens.

There is another condition of neurasthenia which I wish
to place distinctly before you.   It consists in a morbidly
fluctuating blood-supply to the affected parts of the nervous
system.   The nutritive lesion described seems always to
involve an unsteady circulation of blood in the part which
has suffered.   Especially is the blood-supply likely to
fluctuate rapidly, from a too free and tumultuous supply for
healthy nutrition, down to a partial anæmia.   Nutrition in

the part in question is therefore irregular and unreliable. Hence nervous action is unsteady and unreliable. The condition of the walls of the blood-vessels themselves, and of the local vaso-motor mechanisms which control them, in the diseased nervous centres, is much the same as that of the worn nerve elements in the midst of which they are. The local vaso-motor apparatus become worn and exhausted, and the muscular walls of the vessels in the seat of disease lose their tonus in a measure, and give way before any increase in the blood-pressure with abnormal ease, and this is one of the chief conditions of a fluctuating circulation of blood in the diseased nervous centres. When it is known how sensitive are the higher nervous centres to changes in the pressure and quantity of the blood circulating in them, it will not be difficult to understand how rapid and extreme fluctuations in blood-supply may give rise to many symptoms, varying according to the function of the part which is the seat of the disorder.

CHAPTER XX.

If we consider the serum of the blood in the habitual drunkard as an alcoholized fluid, and that alcohol existing in a certain percentage in the serum acts not only upon the serum of the blood but also upon its anatomical elements, we have a condition that modifies nutrition, producing metamorphosis and degeneration of tissue. If, in connection with this, we add the fact that alcohol, *per se*, is a *narcothic* irritant, producing modification as well as degeneration of tissue independent of blood changes, we have an additional reason to regard alcohol as a disease-producing agent.

We have, then, to study the pathological effects produced by alcohol on the blood from two standpoints :

First.—As to its effect on the blood itself.

Second.—The direct effect of the alcohol in the alcoholized blood or serum upon the tissues of the body.

It would be of interest to determine to what extent the serum of the blood can take up alcohol. That it does so, in common with the other fluids of the body, there can be no doubt. Blood taken from an habitual drunkard, and exposed to heat, will give off the fumes of alcohol. At autopsies on drunkards, the fluid in the ventricles of the brain has been ignited with a match.

There is good reason to warrant the conclusion that not only the serum of the blood, but also the fluid of the ventricles and the cerebro-spinal fluid, in the case of habitual drunkards, contain alcohol to a greater or less extent, in some cases probably as much as is compatible with life. A series of chemical analyses, to determine the average percentage of alcohol in the blood of habitual drunkards, would be of extreme interest. The blood in such a condition has not only its nutritive qualities very much impaired and its oxygenation and circulation retarded, but as a builder up of tissue it must be very inferior to normal blood; in fact, it is a disintegrator of tissue. The urine of the habitual drunkard contains a certain percentage of alcohol. The application of heat or the proper chemical tests for alcohol, if applied, prove this. Indeed, if we test the urine of an abstainer within a reasonable time after he has taken alcohol, the chromic acid test will show the characteristic reaction. The effort to prove that the milk of nursing mothers using beer or other alcoholic beverages did not contain alcohol, has resulted in failure. The toxic effect on the infant is shown in the moderate alcohol coma it experiences after nursing, and where the mother was intoxicated, the convulsions that ensued.

The experience of M. Lallemand, Duroy, and Perrin seemed to demonstrate that alcohol received into the body was eliminated by the lungs, the kidneys, and the skin, completely and as alcohol, and that if it was retained in the tissues it was not transformed. The experiments of Anstie, on the other hand, disprove this. " While a certain proportion of the alcohol injested is excreted by the lungs, kidneys, and skin, a certain proportion is broken up in the blood and transformed into some other substance, probably aldehyde," just as aldehyde shortly after its administration is transformed in the blood into acetic acid. But this does not weaken the practical fact that alcohol is present in all the fluids of the body, passes through all the excretory organs, acts directly upon the nervous system and other tissues of the body, producing its deleterious effects either as alcohol or some transformation of it equally pernicious. The degeneration and alteration of tissue in chronic alcoholism is due to the following causes :

First. An impoverished, alcoholized blood, imperfect in its oxygenation and retarded in its circulation, and, consequently, producing mal-nutrition.

Second. The direct irritating effects of alcohol contained in said blood.

Third. The degenerating effects of alcohol on the nervous centers, producing vaso-motor paralysis and impaired reflex action.

The latter is regarded by some writers as the primal and most potent cause of general alcoholic degeneration. The principal tissue changes in chronic alcoholism are fatty, fibroid, and atrophic.

The most marked examples of alcoholic fibrosis are found in the lungs, kidney, and liver.

In the lungs, as alcoholic phthisis, a chronic interstitial pneumonia ; in the liver is cirrhosed, gin, or hobnail liver ; in the kidney as cirrhotic, hard, or contracted kidney. These fibroid changes are slow, may take years to form, but they are rapid in the latter stage. In syphilis we may resolve a gumma or modify the lesions of the tertiary stage. In pulmonary tuberculosis we may be fortunate enough to secure cicatrization of cavities, or hold the disease in check; but the lesions of chronic alcoholism are progressive, and, when once fully established, irremediable, whether in the lungs, kidney, or liver.

The effect of chronic alcoholism on the generative functions in both sexes is instructive and interesting.

"Lippich has demonstrated that alcoholized marriages produce two-thirds less children than among those who were temperate. There can be no doubt that alcoholism affects the generative function of both sexes. The testicles undergo degeneration in alcoholized persons. The spermatic fluid shows this in the well marked changes it exhibits, robbing of the vitality indispensible to conception."

"The alcoholic cachexia after it has attained sufficient

intensity, will produce this, although the organs themselves may not be diseased. Many examples of women are noted who have had children by their first marriage whose subsequent union was barren with an alcoholized husband, and also the reverse. Women may become sterile by alterations of the ovaries and matrix, and abort before maternity. From this point of view alcoholism is a more serious trouble in the mother than in the father."

Drs. Mairet and Combernal recently presented some experiments on the hereditary influence of alcohol before the Academy of Sciences of Paris.

A healthy bitch was made a chronic alcoholic, and gave birth to twelve puppies ; two were still-born, three died by accident, and the remaining seven died of epileptic attacks, enteritis, pulmonary and peritoneal tuberculosis. The lesions found at the post-mortems were thickening of the bones, fatty degeneration of the liver, adhesion of the dura-mater, and other marked alcoholic changes.

A strong bitch was kept intoxicated on absinthe the last three weeks of gestation. Six puppies were born ; three died at birth ; two were of defective intelligence ; one grew up, but was defective in intelligence and nervous organization. This one was coupled with a healthy dog ; of this union three puppies were born ; one died of marasmus, the other two were congenitally defective, having atrophy of hind legs. One of the conclusions drawn was that the degeneration from alcohol was more prominent in the second generation than the first ; also that alcohol used by the mother always produced defective offspring.

A point of interest in this connection is the etiology of dipsomania. The best authorities now agree that, while exceptionally dipsomania may arise from traumatism or alcoholism, the great majority of cases are traced to an insane or intemperate parent or parents. It is a hereditary, not an acquired neurosis.

It will be of interest to record further the results of

chronic alcoholism in the lower animals, produced by experimenters with the view of determining the pathological lesions of alcohol. And none have been more zealous than the French in this direction ; and of these investigators none more prominent or painstaking than M. Magnan.

"M. Tardieu has found meningeal hæmorrhages in persons dying in a state of intoxication. These are less frequent in animals, and this is the reason why pachymeningitis due to the prolonged action of alcohol is rarer in animals than in man."

Magnan.—"That in dogs, even at the end of two months of alcoholic poisoning, the liver undergoes fatty degeneration. A microscopic section shows the cells have lost form, are swollen, round, infiltrated with granules and drops of fat."

M. Pupier notes the effect on a fowl to which absinthe had been given as a drink ten months. "The liver is hard, resistant, lessened in volume, has irregularities on its two surfaces, numerous whitish depressions, the intermediate parts of a reddish brown color. The microscope shows dilatation of vessels at periphery of lobules, filled with granules ; extreme compression and degeneration of hepatic cells."

In another experiment a fowl was subjected to the action of red wine for ten months. "The liver is of a clear yellow color, soft, pasty, and oils the blade of the scalpel. Microscope shows cells enlarged and rounder than normal, filled with granules resembling those in parenchymatous inflammation at its beginning ; here and there large fat drops."

A fowl was given white wine under similar conditions. "The liver is of good color, but is shriveled on its lower surface and borders. Microscope shows dilatation of vessels, which appear three or four times the normal size when compared with the cells which have undergone atrophic degeneration. A rabbit was subjected to alcohol. The liver shows nothing as regards capillary net-work ; the cells are altered and contain two or three nuclei ; around bile ducts there is an increase of connective-tissue nuclei. M. Pupier concludes that absinthe affects primarily the stroma without producing new connective tissue or sclerosis of walls of vessels. This marked new growth has not been confirmed. As for red and white wine and alcohol, their injurious effect is seen rather in the plasma and hepatic parenchyma."

Alcohol would seem to produce hepatic steatosis, but not to the exclusion of sclerosis.

A prolonged period of alcoholic intoxication, and consequent irritation, might provoke sclerosis. In the same animal, with fatty degeneration of liver, are found irritative lesions, such as pachymeningitis, sclerosis of posterior columns of spinal cord, thickening and opacity of arachnoid and pia mater, milky patches in pericardium—all these at the same time.

"The kidneys, like the liver, undergo beginning fatty degeneration. The surface is smooth and even ; the cortical substance and prolongation between the pyramids of Malphigi show a well-marked yellowish tint, with small striations of a deeper color. The microscope shows tubuli, slightly swollen, cloudy, filled with granular and fatty epithelium."

M. Ruge mentions adhesion of capsule to renal substance in four cases ; in three cases fatty degeneration of the heart. Magnan has seen traces of pericarditis : "The coats of the stomach in dogs who take alcohol mixed with food are not sensibly thickened ; but the mucous membrane is injected, rarely ulcerated."

When alcohol is taken without food, and directly injected by œsophageal tube or by fistula, traces of violent gastritis are seen ; in one case the stomach was shriveled and thickened, and the surface of the reddish-brown mucous membrane was lined with a layer of thick, sticky, glairy mucus, streaked with blood. On cleansing with a stream of water, small ulcerations with irregular borders were seen ; in some places cicatrices appeared as irregular grayish plates. In the mucus were found infiltrations of blood, some in layers, others in small spots. In the same dog, the cord is not injected and appears normal ; a grayish tint is seen on the posterior columns, more marked on lower third, where it has the form of a triangle with the base directed backwards on each side of the posterior median fissure ; in the same locality a slight grayish tint in the anterior columns on each side of commissure ; Magan has noted the same condition in a man where chronic alcoholism terminated in general paralysis.

Kremiansky, in dogs who were given alcohol four weeks, noticed pachymeningitis.

M. Neumann observed the same fact, but also that it did not exist sometimes in a more prolonged use of alcohol. Magnan found slight infiltration and slight thickening of arachnoid and pia mater, but no false membranes of dura mater. Others, slight dilatation of vessels of dura or simple injection or œdema of pia.

This diversity is explained by Magnan as due in some cases to a meningeal hæmorrhage during drunkenness, a hæmorrhagic pachymeningitis ; but while this accounts for the existence of new membranes in some dogs, Magnan asserts "that pachymeningitis may come on without pre-existing hæmorrhage in certain nervous affections and chronic alcoholism."

We have thus quoted extensively from these French experimenters that others may be encouraged to follow similar researches with regard to the " pathological effects of alcohol " in animals, with the advantage of improved pathological knowledge and modern appliances ; for these investigations demonstrate not only that researches as to the effects of alcohol can be satisfactorily conducted in the lower animals, but they also corroborate what has been

demonstrated to be the effects of chronic alcoholism on pro-
creation in the human species, as well as its other pathological effects.

Alcohol acts not only indirectly through the blood as an
irritant, provoking fibrosis or other tissue changes, "but on
the alimentary canal, particularly the stomach. The local
effects of habitual doses of concentrated alcohol are seen in
the permanent congestion of the blood vessels, exaggerated
or vitiated secretions from the gastric glands, and ultimately a degenerative change in the structure of the sub-
mucous tissues, which consists in the disappearance of
characteristic secreting structures and hypertrophic exaggeration of fibrous tissues."

The effect of alcohol upon muscular or other tissue, producing fatty degeneration, is similar in this respect to the
action of phosphorus, arsenic, or other poison. Fatty degenerations of the pancreas from alcoholism shows "the
glandular parenchyma has partially or entirely disappeared ;
it may be replaced by adipose tissue, which is developed in
the fibrous stroma of the organ around its vessels and
glandular ducts." In some instances the acini or characteristic gland structure is lost entirely and replaced with
fatty tissue.

But the most marked evidence of the deteriorating
effects of alcohol is seen in its action on the nervous system.
"It is clear that the nervous centers, independently of the
ill effects on their nutrition of the blood changes, have a
certain chemical attraction for alcohol, which accordingly
is found in their tissue.

"The characteristic changes which have been observed
in the brain, medulla oblongata, etc., of confirmed drinkers,
consists essentially of a peculiar atrophic modification by
which the true elements of nervous tissue are partially removed ; the total mass of nervous matter wastes, serous
fluid is effused into the ventricles and the arachnoid, while
simultaneously there is a marked development of fibrous

tissue, granular fat, and other elements which belong to a low order of vitalized products."

From these conditions arise vaso-motor paralysis, with all the results that follow a defective supply of blood and an impaired circulation, tending to local stasis. Moreover, if we exclude traumatism, there is not any disease of the nervous system resulting from other causes than alcoholism, at least with few exceptions, that alcohol cannot produce—alcoholic neuritis, alcoholic anæsthesia, general paralysis, serous apoplexy, etc., and those cerebral conditions from which arise the acute and chronic forms of mental derangement. The nerves of special sense are not exempt. "The abuse of alcoholic stimulants has been said to be the cause of amaurosis, and, as a proof of this, the fact has been adduced that the affection has been arrested, or even cured, by completely giving up the habit of drinking. This much is certain, that amblyopia occurs in great misproportion among habitual drinkers. It is generally first seen as night-blindness, but soon becomes constant, and gray atrophy of the nerve is recognized by the ophthalmoscope." (Pagenstecher.)

"One point of interest in this connection, relating to the action of alcohol on the nervous system, is the theory advanced by writers on this subject, and it is a very plausible one : that the degeneration of all tissue in cases of alcoholism is due primarily to the action of alcohol on the nervous centers, and through these, by vaso-motor disturbance or impaired reflex action, upon the organs or tissues which these nerve centers, or vessels influenced by them, supply." But it would seem, while regarding this as the prime cause of alcoholic degeneration, we could not ignore the fact that the blood itself was a chronic alcoholism much deteriorated as to its quality and retarded as to its circulation, and, moreover, that it contained a chemical irritant. The limits of this paper will not permit us to consider in detail all the pathological changes due to alcohol. It affects all the tissues of the body ; even the bones are not exempt.

.   We have therefore generalized our statements and taken a view over the whole field, rather than endeavored to carry out and elaborate any special line of thought.   Our object has been to demonstrate that there is abundant material for the pathologist and the microscopist to investigate, and a neglected but nevertheless a rich field for medical research. How little progress has been made in the study of the pathology of chronic alcoholism and the diseases incident to alcoholism.

A general principle may be laid down that, whatever tends to innutrition aims directly at that strength and balance of its forces, that coördination, so to speak, between its peripheral and central portions, that it is needful for the equable discharge of its multitudinous functions.   Innutrition, by lowering the vitality of the brain-cells, and diminishing the store of power held by the central ganglia from the steady and well-timed responses to all the demands upon them, into spasmodic, irregular, and insufficient supplies of the force which it is their province to furnish.   But alcohol especially promotes innutrition, and the very stimulation which it produces is the surest evidence of its drain upon those reserve forces, that exuberance of the central nervous fund, that wealth of power which is indispensable to the maintenance of the full vigor of the constitution during those brief and rare occasions when unforeseen circumstances shall make unusual demands upon them.

Nor is this exhaustion and innutrition all the evil which alcohol works in the constitution.   The blood and secretions are vitiated and loaded with material foreign to their normal constitution, and there is a universal departure from that almost infinite delicacy of balance, resiliency of the organization, which in the natural healthy state characterizes its various portions, to say nothing of that deprivation of the higher spiritual nature which is the inevitable concomitant of the habitual deviation from natural methods which is forced upon it. Nor is this all of the evil. How unreasonable it is to suppose that children begotten of a parent during such exhaustion of the ganglionic force—during such prolonged vitiation of the blood and secretions and the perversion of the intellectual and moral forces—should not carry in their physical and spiritual natures evidence of the outrage done to natural laws !

CHAPTER XXI.

GENERAL CONSIDERATIONS OF TREATMENT, PROGNOSES, RESULTS, ETC.

Four conditions of cure must be observed. The first condition of cure and reformation is abstinence. The patient is being poisoned and the poisoning must be stopped. Were it an arsenic instead of an alcohol no one would dispute this ; so long as the drinking of intoxicants is indulged in, so long will the bodily, mental, and moral mischief be intensified and made permanent. Abstinence must be absolute, and on no plea of fashion, of physic, or of religion, ought the smallest quantity of an intoxicant be put to the lips of the alcoholic slave. Alcohol is a material chemical narcotic poison, and a mere slip has, even in the most solemn circumstances, been known to relight in the fiercest intensity the drink crave which for a long period of years had been dormant and unfelt. The second condition of cure is to ascertain the predisposing and exciting causes of inebriety, and to endeavor to remove these causes, which may lie in some remote or deep-seated physical ailment. The third condition of cure is to restore the physical and mental tone. This can be done by appropriate medical treatment, by fresh air and exercise, by nourishing and digestible food given to reconstruct healthy bodily tissue and brain cell, aided by intellectual, educational, and other influences. Nowhere can these conditions of cure be so effectually carried out as in an asylum where the unfortunate victim of drink is placed in quarantine, treated with suitable remedies until the alcohol is removed from his system, then surrounded by elevating influences, fed with a nourishing and suitable diet, and supplied with skillful medical treatment. His brain and nervous system will then be gradually restored to its normal condition, and, after a period of from six to twelve months in most cases, he will be so far recovered as to be able to return to his usual avocation and successfully resist his craving for drink.

In the cure of inebriety there is probably more agitation and interest than ever before. The efforts of societies and parties, of the pulpit and rostrum, with the increasing books and papers from the press, have never been more active than to-day. Yet reports show that inebriety is increasing, and that more spirits are made and consumed every year.

All the temperance efforts and legal means for the cure and prevention of inebriety are based on the theory that it is a moral disorder which the victim can control at will, or a wicked habit that he can continue or put away at his own pleasure. This theory of inebriety is theoretical, and embodies the same error which follows every new advance of thought, namely, explaining all human action from some moral or theological standpoint. Thus the phenomenon of insanity was explained as a possession of the Devil, and the victims were supposed to enter into a compact with evil spirits, voluntarily. The remedy was severe punishment. Public attention was occupied for ages in persecuting and punishing the insane and epileptics on this theory of the causation. Law, religion, government, and public sentiment, all failed in the cure and prevention by this means, and these diseases went on unchecked, simply because the real causes were unknown.

Inebriety is regarded in the same way as wickedness, and the same means are urged as a remedy. Over five hundred thousand inebriates were sent to jail in 1891, and punished as willful and voluntary drunkards. Armies of moralists and temperance people are pledging and praying the inebriate to stop drinking, and exercise his will, and be temperate and well again.

Yet all such efforts fail, and often tend to increase the very condition which they seek to remedy. They fail because they are based on a false assumption of the causes, and not on any accurate study of the history or real condition of the patient. A new era is dawning for the inebriate. His diseased condition, and the need of special medical care in special surroundings, is a truth that is spreading slowly and surely in all directions. Not far away in the future inebriety will be regarded as small-pox cases are now in every community. The inebriate will be forced to go into quarantine and be treated for his malady until he recovers. The delusion that he can stop at will, because he says so, will pass away. Public sentiment will not permit the victim to grow into chronic stages ; the army of moderate and periodic drinkers will be forced to disappear, and the saloons which they have supported will close in obedience to a higher law than any prohibition sentiment.

Public sentiment will realize that every inebriate is not only diseased, but dangerous to society, to himself, and all his surroundings, and demand legal guardianship and restriction of personal liberty until he recovers. When these poor victims realize that society will not tolerate their presence or allow them personal liberty in this state, they will seek help and aid before they reach extreme stages.

This is the teaching of all modern science,—to check the disease at the beginning, to seize the poor waif on the street and the rich man's son, who are just at the beginning of inebriety, and force them into conditions of health and sobriety, to save the one from becoming a prey on society and a burden to the producer and tax-payer, and the other from destroying society and himself and leaving a tide of misery and sorrow that will continue long after. When society shall realize and act on these facts, the great centers of pauperism and criminality will be broken up. This will be accomplished by the establishment of work-house hospitals, where the inebriate can be treated and restrained. Such places must be located in the country, removed from large cities and towns, and conducted on a military basis. They must have all the best appliances and remedial means to build up and restore the debilitated victim. They should be military training hospitals, where all the surroundings are under the exact care of the physician, and every condition of life is regulated with steady uniformity.

Besides the medicinal and hygienic treatment, there should be educational and industrial training, and each one should be employed, both in body and mind, every day. He should be placed in a condition for the best culture and building up of the entire man. Every defect of body and mind should be antagonized and remedied as far as possible. Each case should be an object of study to ascertain the real state and the means to strengthen and improve it. These hospitals should be built and conducted entirely from the license fund or the taxes on the sale of spirits. They should, in a large measure, be self-supporting from the labor of the inmates, and independent of the tax-payers. These places would most naturally divide into three distinct grades. The first class of hospitals should be for recent cases, where the inmates can be committed by the courts, or voluntarily commit themselves for one or two years. The second class should receive chronic cases for longer terms of treatment —from one to three years. The third class should be for the incurables, or those who give no reasonable promise of restoration. The time should be from five to ten years and life.

The latter class should be thoroughly organized into military habits of life and work, and kept in the best conditions of forced healthy living. Employment and mental occupation should be carried out literally as a stimulus to strengthen the body and mind. Where it was possible the rewards of his labor, beyond a sum to pay for care, should be turned over to his family and friends or held in trust for him.   He should be encouraged to healthy work and living by all possible means and surroundings.   The semi-chronic cases should be treated substantially the same way, only occupation and training of the mind and body should be more suited to the wants of each case.   The amusements should also be of a sanitary character.

The recent cases should have the same exact discipline, filling the mind with new duties and new thoughts, and suited to build up the exhausted, overworked man, as well as the gormand and under-worked idler.   All persons should pay for their care if possible, and be required to render some service which would be credited on their bills. These hospitals should be literally quarantine stations, where the inebriate can be housed and protected and society saved from the losses following his career.

If ten thousand poor chronic inebriates could be taken from New York and placed in such hospitals, and made self-supporting, who could estimate the gain to society, to morals, to the tax-payer, and to civilization ?   This can and will be done in the near future.   If ten thousand semi-chronic cases of inebriety could be taken from New York and quarantined two or five years in such military hospitals, and made to pay for their care by labor, who could estimate how many would be returned to health and temperate living again ?—who could estimate the relief from sorrow, misery, wretchedness, and losses ?   This will also be a reality a little farther on.   If ten thousand recent cases of inebriety could be taken out of their surroundings in New York and placed in these hospitals, where forced conditions of the highest degree of health and vigor are maintained, a large percentage would recover.   The gain to society and the world would be beyond all computation.   Now each one of these propositions and the practical working of a military hospital is a reality, based on evidence constantly accumulating.   Every prison, penitentiary, or hospital, every

asylum or home where inebriates come under care and
restraint bring such evidence. They show that such a
method of treatment, combining the varied experiences of
all these institutions can be made practical and is the only
scientific way of solving this problem.

To banish the still and saloon does not prevent inebriety
or cure the inebriate ; it only changes the direction of the
drink current. But quarantine the inebriate in a hospital,
as one suffering from contagious disease, and the victim is
cured, the spread of the disease is prevented, and a knowledge of the causes ascertained, from which the remedies
can be known and applied. To punish the inebriate as a
criminal cannot cure his inebriety, but it always unfits him
for living a temperate, healthy life hereafter. To attempt
a cure by faith and prayer is to depend on false hopes, the
failure of which is followed by increased degeneration.
To attempt any form of treatment without knowing any
other fact except that the victim drinks to excess is always
to blunder and fail.

The time has come to recognize the physical conditions
which enter into all cases of inebriety, and to apply exact
remedies along the line of nature's laws and forces.

The late Dr. Bellows, in an address delivered in 1858,
said : "Inebriates, like criminals and insane, will all be
eventually restrained in hospitals, and treated with medical
and psychological skill the moment their liberty becomes
dangerous to society. The terms of their confinement will
be limited only by the possibilities of cure and the conditions of their disorder. Society gains nothing by holding
prisoner for an hour any man who is fit to be at large.
Liberty and human rights gain nothing by allowing any
man to be at large for a moment who is destroying himself,
his family and his neighbors. What we need is what we
are fast gaining, namely, a possession of the tests and
gauges of the fitness and unfitness, and we shall be able to

treat the inebriate successfully the same as in other diseases."

Some general idea of the details of treatment will be of
interest. In a hospital conducted on scientific common-
sense principles the patients are received for periods of not
less than three or six months. The patient signs a commitment paper, and is examined by the physician, and all the
facts of his present and past condition noted. If intoxicated he is placed in charge of a nurse and baths and
remedies given for his special condition. If sober he is
given a pleasant room and placed upon a regular diet, exact
conditions of living, and required to take such medicines,
baths, exercise, and general treatment as may be needed in
his case. Mental occupation, amusement, change of thought
and life in every particular are sought for. He is treated
as one who has a profound disease of the brain and nervous
system, requiring rest, care, and removal from every source
of irritation and excitement. The question of responsibility
to aid the efforts of the hospital in his behalf is urged as a
symptom of his capacity or incapacity to recover. The
asylum is a quarantine where he can recover, and his
liberty or restraint is governed by his condition. Wherever
congenial work can be added to the amusement it is done
as a medical aid. Every condition of life is controlled and
regulated, and every surrounding arranged to aid recovery.
Daily religious exercises, rides, walks, Turkish baths, and
exact methodical living, most naturally result in a degree
of strength and vigor that is very promising for the future.

To accomplish this the asylum must be pleasantly located,
the building must be cheerful and sunny, and adapted to
the work. The surroundings must be free from all sources
of irritation and unpleasantness, such as noise, excitement ;
and the management must seek to apply remedies to meet
the wants of every case. Thus, in some instances, sharp
restraint is a tonic remedy of value ; in others it is depressing. In some cases nerve rest and regularity of living is of

more value than all other things. The physician finds
from a study of each case a most bewildering complexity
of causes, beginning with heredity and including everything which produces distinct strain or drain on the nerves
and brain.

The great object of all inebriate asylums is to practically quarantine the inebriate, where all the exciting causes
are removed, and every predisposing cause is antagonized,
and where the most favorable conditions for rest and building up can be secured. The remedies to aid in this
work are indicated by the special condition present, and
include all the nerve and brain tonics, the common property
of the profession. The physical remedies, are diet, baths,
exercise ; and the mental remedies, including mental occupation, amusement, change of thought and mind, growth
out of the past, are all essential and equally valuable.
Restraint and freedom, rewards and punishment, appealing
to the will and ignoring it, etc. Rules and regulations for
every act of life ; few rules, which the patient applies himself, and so on, using the most diverse means which would
apply to the special need of each case.

The condition or disease to be treated is a profound
physical disorder, and the toxic use of alcohol is always a
symptom as well as a cause. The best efforts of science is
to break up this cause, and change the symptoms. The
defects continue, the man is permanently impaired. He may
go back to active life and do well, but ever after, as a rule,
the same range of causes will produce the results.

In a certain number of cases the drink impulse or
symptom seems to be permanently exhausted after a time,
like the exhaustion of the germ soil of some diseases,
and no exciting causes will develop the drink symptoms
again. Other defects may appear, but he never again uses
spirits. The germ soil has gone, it may be forever, or after
lapse of years it will return.

A period of six or twelve months in an asylum will remove the states of delirium which have kept up the use of alcohol, and reveal an exhausted brain and nerve soil that will not tolerate alcohol in any form after. The person suddenly realizes that alcohol is both poisonous and repelling to his system. This may be so intense that should he take any form of spirits by mistake it will produce intense nausea and depression. In other cases the drink exhaustion dies out after long years of abstinence, and should the patient relapse late in life, death follows soon after. A careful study in an asylum often reveals these cases, and the expert can safely predict a total or partial cure of the drink symptoms and disorder, or its temporary suppression, only to break out again. Another fact, not generally known, appears to the asylum physician, namely the great uniformity of the symptomology and progress of these cases. Beginning at a certain point, or from certain range of causes, they follow a uniform line of progress, which can often be seen, traced, and predicted with certainty.

In an asylum the stage of progress can frequently be seen, and means for its prevention or diversion can be applied. In an asylum all the conditions of life can be regulated ; the food, the surroundings, the period of rest and activity ; the mental states can be antagonized or prevented ; the local condition from which irritation is produced can be removed ; the physical irritation which keeps up the drink impulse is changed. The asylum treatment, like the quarantine for contagious disease, isolates the victims from all exciting and predisposing causes, and thus places him in the best possible condition for returning health. Asylum and medical treatment for the inebriate are most imperatively demanded, not only to save the victim, but to enable us to understand some of the great underlying causes which are active in precipitating so many men into this terrible disease.

Primarily in the treatment we have shattered constitutions, and broken-down nervous systems to deal with. We have a disease eminently marked by weakening of the will-power, and seclusion from society, rest, judicious restraint, and enforced abstinence from all alcohol stimuli are cardinal points of treatment. I always let patients, applying to me for treatment, distinctly understand that a permanent recovery depends largely on allowing sufficient time for restoration of nerve-power, mental tone, and physical vigor, and I think, in most cases, six months is the least time necessary for a complete recuperation of the will-power. Dipsomania is a disease that requires the most perfect discipline, both moral and physical, if we expect a cure. Periodical insanities are notably difficult to cure so that there is no chance of a relapse, but we may reasonably expect an ultimate cure if there is but little structural change in the brain which has resulted from the course of inebriety.

The heart's action in dipsomaniacs is weak, often irregular, accompanied by palpitation. There is a loss of tone in character, blunting of moral perceptions, impairment of intellectual discrimination, and generally impairment of all the mental faculties. There is very often an utter inability to fix the mind on any one subject, or to follow up a train of thought consecutively. We see periods of abnormal cerebration ; an instantaneous abeyance of reason and judgment ; a condition resembling an epileptic state, during which period the inebriate may be actuated by mad, ungovernable, or eccentric impulses ; a condition in which disease has deprived him of the power of choice, and during which states, in my opinion, the inebriate is as little responsible for his actions as is a person suffering from any other phase of mental disease. I knew a gentleman of high social standing in this city who has these mental blanks, during which time he rides up and down New York in a horse-car, aimlessly, and has no recollection afterward of having done this at all.

These are very interesting cases from a medico-legal point of view. In these cases the healthy coördination of ideas is destroyed for the time, and the patient need not be alcoholized to have such states occur. In cases of inebriety or dipsomania we have to deal with the results of a toxic poison, and we may have various complications proceeding from the abuse of alcohol, such as cirrhosis of the liver, gastritis, epilepsy, various forms of dyspepsia, and in some cases with Bright's disease. We may also find a simple hypertrophy of the left ventricle of the heart, and degenerated arteries. Our patient must have cheerful, tranquil, and pleasant surroundings ; all cerebral excitement must be repressed ; sleep must be procured ; plenty of nourishing, easily digested food administered at short intervals, and an abundance of fresh air and exercise. I am not in favor of much meat in the diet list. Certainly none at all, except at dinner, and I think a strict diet list, so that the work of the liver and kidneys is diminished by lessening the amount of highly nitrogenized food and of the hydro-carbons, gives decidedly the best result. Fish, oysters, and fruits that contain sugar, the green vegetables, bread, wheaten meal, and oatmeal are all permissible. Potatoes should be eaten somewhat sparingly, and eggs also. Poultry and the white meats are well borne. Beef and mutton are not so well borne, and are contra-indicated, except at dinner.

Exercise in the open air is indispensable. The action of the skin must be stimulated, and systematic skin friction with cold sponge or shower baths do much good. Also the Turkish bath is useful. The natural saline mineral waters are the best to stimulate excretory action by the abdominal organs. The Hunyadi-Janos, Friedrichshall water, and Saratoga Congress water are the best, while for diuretic waters, Saratoga Vichy and Apollinaris water are the best.

We have a worn, irritable condition of the nervous system in inebriates, an unstable condition as regards its nutrition, its solidity, and its perfection of structure. We must supply the greatest amount of nutritive material to the brain and nervous system to repair the existing nutritive lesion. We must supply the phosphates in some way, the many eligible preparations containing them being familiar to you all. All abnormal nervous excitability must be quieted, and our patient kept calm and tranquil. A prolonged warm bath at night, followed by the administration of one teaspoonful of the ammoniated tincture of lupuline, or of a pill of camphor, ext. hyoscyamus, and pulv. digitalis will often quiet the nervous excitability.

Care must be given to the excretory functions of the skin, kidneys, and bowels. If there is headache and drowsiness, such diuretics as the liq. ammoniae acetat:, with spt. nitric ether are indicated. Indian hemp has proved itself in my hands a valuable adjunct, in doses of ¼ gr. of the extract, as required. A very valuable sedative mixture is one composed of 30 grains of bromide of sodium and 30 minims of tinct. of cannabis indica prepared with water, at the time required. It induces calmness and tranquility, and can be repeated thrice daily, if required, without the loss either of flesh or appetite. One of our most valuable remedial agents in inebriety is phosphorous, which I always give in 1-100 to 1-25 gr. doses in cod liver oil after meals. Cod liver oil is one of the best nutritive remedies, as fat must be applied to the nutrition of the nervous system if this is to be maintained in its organic integrity. The general effects of phosphorous are those of a stimulant, but it possesses a special power over the exhausted nervous system. It is perhaps evanescent in its effects, but is never followed by a stage of depression which is noticable. It should never be ordered upon an empty stomach. A pill of iron, phosphorous, zinc, and strychnia is also very useful, and combinations are often more beneficial than the medicines when taken singly.

A large percentage of all cases of inebriety, before admission into our institutions for the cure of inebriety, have passed through the primary and acute stages, and have probably been subjected to medical treatment. This fact must never be lost sight of in forming our opinion, not only of the nature of the disease itself, but of the medical treatment necessary for its cure. We often discover that the dipsomania has been allowed to exist and slowly progress for a considerable period, no treatment, either medical or moral, having been adopted for its removal.

The most simple classification of inebriety, the one best adapted for useful and practical purposes, is its division into the *acute, periodic,* and *chronic* forms. The *acute* form is the rarest, and is ushered in by exhausting diseases, or excessive sexual excess. The *periodic* form is much more frequent, and it is to this form of inebriety that I would recommend the term *dipsomania* to be restricted. It is frequently hereditary, like all insanities, and strictly periodic in its paroxysms. These patients—dipsomaniacs—may abstain for weeks or months, and may during this interval positively dislike stimulants.

The last or *chronic* form of inebriety is very incurable, as the patients are incessantly under the desire for alcoholic stimulants, and will get them if any opportunity occurs, whereas the *dipsomaniac* have only the irresistible craving for stimulants *periodically.* It is in these cases of *chronic inebriety* that we find hallucinations of sight and hearing, very painful moral impressions, confusion of thought, suicidal tendencies, tremors of the facial muscles and tongue, and very often paralytic symptoms, ending in general paralysis. It is here, in chronic inebriety, that disease of the brain may destroy all apparent consciousness of pain, and keep in abeyance the outward and appreciable manifestations of the important indications of organic mischief. Extensive disease of the stomach, lungs, kidneys, bowels, uterus and heart have been known to have progressed to a fearful extent without any obvious recognizable indication of the existence of such affections.

The most essential preliminary matters for inquiry

relating to the treatment of inebriety, have relation to the
age, temperament, previous occupation and condition in
life of the patient.   It will be necessary to ascertain the
character and duration of the attack, to ascertain whether
it has resulted from moral or physical causes ; if of sudden,
insidious, or slow growth ; whether it has an hereditary
action, or is the effect of a mental shock or of mechanical
injury ; whether it is the first attack, and if not, in what
features it differs from previous paroxysms.   It will also be
our duty to inquire whether it is complicated with epilepsy,
insanity, suicidal or homicidal impulses.   If any prior treatment has been adopted we must learn its nature ; whether
the patient has suffered from gout, heart disease, rheumatism, skin disease, or syphilis.   It is important in cases of
females to obtain accurate information in relation to the
condition of the uterine functions.   We should also inquire
whether the patient has been suspected of habits of self-
abuse.

Having obtained accurate information upon these essential points, our own personal observation will aid us in
ascertaining the character of the inebriety.   The configura-
tion of the head, chest, and abdomen ; the gait of the patient ; the degree of sensibility and volitional power manifest ; the state of the retina, the pulse, the temperature of
the head and body ; the condition of the skin and chylopoie-
tic viscera ; the action of the heart, lungs, and nature of any
existing disease of the uterus should be noticed.   Our *prognosis* in cases of inebriety will mainly depend upon the duration of the attack, its character and origin, and the diathesis of the patient.   The prognosis is unfavorable if the
disease is hereditary.   Age materially guides us in forming
a correct prognosis.   The greater proportion recover between the ages of twenty and thirty-five.   When a patient
has youth and a good constitution, and remedial measures
are promptly applied to the patient while he is secluded

from all occasion of temptation, the prognosis is favorable. I have seen patients after forty and fifty recover.

The prognosis is unfavorable when inebriety is associated with organic disease of the heart or lungs, or when great impairment of mind, associated with paralysis, is present. Prolonged hot baths are of the utmost service in the treatment of inebriety. Among the therepeutic effects of these baths I would mention a diminution of the circulation and respiration, relaxation of the skin, alleviation of thirst, the introduction of a good deal of water into the system, an abundant discharge of limpid urine, a tendency to sleep, and a state of repose. It is most useful in acute and chronic inebriety. The preparations of hyoscyamus, conium, stramonium, camphor, hops, aconite, ether, chloroform, and Indian hemp are all of great service if given with judgment. The best plan in practice would seem to be to judiciously combine various kinds of sedatives. Milk heated almost to boiling is very valuable. It allays irritability of the stomach and craving for stimuli, and two glasses at night have a very sedative effect.

If there are decided signs of cerebral congestion the occasional application of a leech behind the ear is good practice. If symptoms of softening of the brain appear they will often yield to the persevering use of the preparations of iron, phosphorous, zinc, and strychnia, with generous living. Strychnia antidotes, it appears to me, slight impairment of the mind, loss of memory, defective power of attention in inebriety very perfectly, and may be combined with two gr. doses of quinia thrice daily. If inebriety is associated with suicidal tendencies it will be important to ascertain whether any cerebral congestion exists, and if so, a few leeches applied to the head, followed by an active cathartic, will relieve the local irritation. In the absence of any positive active cerebral symptoms, the prolonged warm bath and the continued exhibition of morphia will be the best treatment until the suicidal idea disappears.

We need both State and national *hospitals for the cure of inebriety*, places of detention or asylums, under the care of medical officers well trained by preliminary education for their vocation, and acquainted thoroughly with inebriety.

One of the best remedial agents which can be employed in inebriety is electricity. Both the constant and induced currents, or galvanic and Faradic electricity, may be used. Electricity is a true nerve tonic, and it is an agent which furnishes us with the means of modifying the nutritive condition of parts deeply situated, and of modifying the circulation to a greater extent, I think, than by any known agent. By the judicious employment of the constant and induced currents we have it in our power to hasten the processes of nerve growth and nerve repair, and thereby hasten the acquisition of nerve power.

The use of electrictity does not, I think, act by contributing anything directly to the growth or repair of nerve tissue. Its action is to stimulate and quicken those processes on which the material and functional integrity of the nervous system depends. The use of electricity in inebriates is always followed by an increase of strength and nerve force, and the results gained are gradual and permanent. Dipsomania or periodical inebriety is characterized by abnormal nervous excitability, conjoined with cerebral exhaustion, and the two indications which are urgent are, primarily, for increased rapidity and effectiveness as regards the process of nerve nutrition, and secondarily, to secure freedom from excitement, and diminution of nerve activity, and thereby to check the waste of nerve structure and of power. These indications we can fulfill by the judicious use of electricity and nerve tonics more certainly than by any other means, there being no other such combined sedative, restorative, and refreshant to the central nervous system. To give the brain the direct nutriment it needs in inebriety, I have before stated, can be accomplished by rest, cod liver oil, phosphorus, the phosphates, etc.

Every intelligent person who has observed the march of human thought into the recently explored realms of Psychological medicine and the treatment of mental diseases, will get new views as to modes of treatment, more especially of that class for whom institutions were organized. The favorable results of each succeeding year, only confirm and demonstrate the truth of the humane and wise idea that led to the organization of institutions, viz.: that intemperance in all its stages may be not only checked and mitigated, but in many instances permanently cured, and the subject fully restored to his normal condition of health and sobriety.

Such results may not be reached by the final and utter extinction of the morbid desire for alcohol, so much as by a development and cultivation of opposite and ennobling qualities, which by their vital action, hold the depraved mental tendencies of the subject in constant and absolute subjection, so that they become as inoperative as if they did not exist.

This is as near an absolute cure as we can hope to reach, as the testimony of all inebriates concurs in the fact that the appetite for intoxicating drink never dies, though it may be put to a life-long sleep.

It cannot be expected that the final and complete results of the treatment of our patients, so variously circumstanced and conditioned, can be fully known. From this common center of reform, hundreds have struck out in new and divergent paths, and are lost to our view in the general whirl of business and laudable enterprise. Whenever any one does fall into his previous habits, we are certain to be informed of the fact, as few things travel so fast and so sure as ill-tidings of man's vices and misfortunes.

We can congratulate ourselves on the fact that thousands who have been under hospital care are now in active life all over the country, of whose doings we are cognizant, and it is a source of pleasure to us to know that their correct and consistent conduct is productive of happiness to themselves, and does honor to the institutions and the humane work which it has in hand.

CHAPTER XXII.

ASYLUMS AND THEIR WORK IN THE TREATMENT OF INEBRIETY.

Asylums and hospitals for the treatment and cure of inebriates are only modern applications of truths asserted centuries ago. Ulpian, the Roman jurist, in the second century of the Christian era, urged the necessity of treating inebriates as sick and diseased, in special surroundings, with special means. Other authorities indorsed these views, and asserted that the State should recognize the veritable madness of drunkards and treat them as such.

In 1747 Condillac, of France, wrote that the State should provide special hospitals for drink maniacs, and urged a change of law and public sentiment to this end. Dr. Rush of Philadelphia, in 1790 ; Dr. Cabanis of Paris, in 1802 ; Prof. Platner of Leipsic, in 1809 ; Salvator of Moscow, in 1817 ; Esquirol of France, in 1818 ; Buhl Cramner of Berlin, in 1822, all urged the need of physical restraint and treatment of the inebriate as sick and diseased, in places especially provided for this class. In 1830 the Connecticut Medical Society appointed a committee to report on the need of an asylum for the medical treatment of inebriates. In 1833 Dr. Woodward, of the Worcester Insane Asylum, in Massachusetts, urged that inebriety be recognized as a disease, and special hospitals be provided for its treatment. In 1844 the English Lunacy Commission urged that inebriates be regarded as insane, and sent to asylums for special treatment. These are only a few of the more prominent references to inebriate asylums, although many other

215

writers urged the same views in different ways. The mention of the disease of inebriety roused a bitter opposition, and the question of asylums was put aside until the former could be settled.

In 1846 Dr. Turner began an enthusiastic agitation, which culminated in the first asylum at Binghamton, New York.

The opposition to this work was very intense, and came from moralists, who urged that it was a purely "infidel work" to diminish human responsibility. The asylum at Binghamton began on the most advanced principles, of receiving no one for less than one year, and having absolute restraint over them during this time. It asked no pledges or promises from the patient; it aimed to give each one positive protection and medical treatment. The patients were locked in at night, and only allowed out under the strict care of attendants. Each case was considered a suicidal case of insanity, requiring long medical care and restraint.

These methods were far in advance of that time, and even to-day are just beginning to be recognized as the latest teachings of science. The patients themselves, after the immediate recovery from the effects of spirits, protested against the confinement and doctrine of disease, and sought in every way through their friends to break up the methods of treatment. The points of difference were these : The asylum and management insisted that each case was more or less diseased, and should be under absolute control and restraint for long enough time to effect a permament cure. The patients and their friends insisted that, while the case might be diseased, his recovery depended largely upon his liberty and promise to get well ; that restraint was irritation and injury ; and that appeals to his honor and manhood were the real agents for final cure. In brief, one plan proposed long restrain ; the other, no restraint except nominal care at first, then persuasion and advice.

The board of management changed, and the central object was to make the asylum popular with the patient. Political controversy followed, and the State changed the asylum to an insane hospital, and a political governor, in justification of the act, called the inebriate asylum a failure.

The asylums in this country for the treatment of inebriates may be divided into three classes.

1st. Those hospitals which have been established by corporations receiving State aid, or by private enterprise, and incorporated, having the advantage of the laws, where the inebriate is regarded as diseased, and treated on broad scientific principles.

2d. Hospitals, both corporate and private, where the inebriate is admitted with other cases of mental disease ; mild maniacs, and eccentrics of all degrees of debility, of both mind and body. Here the inebriate is treated from some mixed theory or half disease and half vice ; when he has recovered from the acute symptoms, is held as fully sane and responsible.

The *third class* of hospitals ignore all question of disease, except that which comes from the direct use of alcohol, which from their theory can be quickly remedied. The treatment of the case is by the pledge, prayer and promise.

The first class of hospitals endeavor to apply real scientific means of treatment and study of these cases, much the same as in other cases of insanity.

The second class combine and treat these cases with other mental diseases, without any special study or recognition of their nature or character.

The first persons who come to these asylums are the incurables. They clamor most importunately for help. They are the credulous, emotional incurables, who have signed pledges, joined churches, and tried every means known, and now expect from the asylum some miraculous power of restoration. In a few weeks they believe themselves fully recovered, and go away only to relapse again and become bitter enemies of the institution. This class appear everywhere as examples of the failure of the hospitals. Another class of incurables come from the better ranks of society, often for the purpose of accomplishing some object, consent to go anywhere and do anything that promises relief or restoration. These moral paralytic inebriates rouse the highest expectation and greatest enthusiasm in the grand work of asylum-cure among their friends, and pose as examples of "brands rescued from the burning," then suddenly relapse and condemn the asylum and management as the cause. The humbug of the asylum, its frauds and deceptions, are themes of great relish and pleasure to them.

A second class of persons are coming in greater numbers every year to these asylums. They are the curable cases,—the nerve and brain-exhausted men and women, the large and ever-increasing class of business and professional men, who have broken down from overwork, worry, and irregularity of life and living, and who find alcohol a narcotic of most seductive nature. The still larger class seen in every city of the land, who from brain strains and drains incident to the rushing, grinding civilization of to-day, also to the struggle for position, wealth and power, and the effort to adapt themselves to the new conditions of life, to the new demands, thus prepare the soil by exhaustion, and encourage the growth of inebriety and its allied diseases. This class often represents the highest talent and genius, and, as a rule, are the active brain-workers of the times. An inebriate hospital to this class is almost an "el dorado." It brings rest, restraint, seclusion, building-up, and is literally a place for repairs and restoration.

In this class the use of alcohol, opium, or any other narcotic is often more of a symptom of exhaustion and debility, for which rest and medical care are essential. There are many thousands of this class who could be saved and permanently restored to temperate life and living if they could be placed in inebriate hospitals and treated early. Later, they become chronic and incurable, and are ever after a burden and heavy loss to the world. To-day they cannot go to an insane asylum, and the public hospital is unfit for them, and no place is open adapted to reach their wants.

A large army of these curable cases are scattered in every community, and in almost every home; and are the skeletons which haunt and peril the peace of many households. They are the secret and moderate drinkers. They are those who have secret or open drink paroxysms, and who recover only to relapse again with steady increasing frequency. Both men and women in all circles of life are found in this army of dissolution. Moral remedies fail, religion fails, they go steadily down and soon all fears of publicity are thrown aside, and the march to death is rapid and distinct. Not far away in the future, asylums and hospitals will be opened to save these cases, and public sentiment will demand that they be placed under treatment early in the progress of the case. Of this class a very large number are curable, and all are benefited by hospital residence.

A third class comprises the erratic border liners, or persons who alternate up and down the line of sanity and insanity, whose genius attracts by its glitter, and bewilders by its weakness. They sound the praises of the asylum far and near, exaggerate its power, and claim the most extraordinary results, then rush to the other extreme on relapsing, exhibiting a malice and pleasure in destroying what they so lately praised.

Others who are less incurable appear, but always wish to decide the length of their treatment, then go away only to relapse and attribute the failure to the asylum. In addition to all this, public sentiment often gives credit to these statements of incurables, and hence withholds the sympathy and aid which should be given. In many cases the State refuses to give only limited authority to the managers to hold patients. And in other cases the clergy and temperance reformers insist that prayer, conversion, and the pledge shall be made prominent in the treatment.

Thus the most extraordinary misrepresentations, extravagant credulity, and ignorant criticism follows every movement of the institution. The superintendent and the managers are never able to carry out their plans fully, or bring out the real object and methods of obtaining it with these cases.

The results of treatment in a few scientific hospitals for inebriates are most encouraging. The first statistical study was made at Binghamton Asylum in 1874. The object was to find out how many persons who had been under treatment continued temperate years after. Accordingly, over a thousand circular letters were addressed to friends of patients who had been under treatment five years before, asking the present condition of the patient. The answers indicated sixty-two and a-half per cent. as yet temperate and total abstainers. This result, after an interval of five years, was clear evidence that a large per cent. would remain cured during the remainder of life.

Dr. Day, of Washingtonian Home, made a similar study of eight thousand cases who had been under treatment ten to eighteen years before, and found over thirty-eight per cent. yet sober and temperate. Dr. Mason, of King's County Home, examined two thousand cases who had been away from the asylum for ten years, and found thirty-six per cent. of all cases yet cured. Other observers have made studies of a smaller number of cases with similar results.

Not less than two thousand inebriates are under treatment in hospitals in America. Over a thousand of this number are in special hospitals. They represent most largely the incurable cases; persons who have tried every means found in the pledge, prayer, and by moral suasion, and exhausted every resource of home and friends, and come as a last resort, expecting extraordinary change and cure. They have been victims of this disease from five to thirty years, and present the most complex and varied degrees of physical and mental degeneration. Yet, notwithstanding this fact, the experience of the few scientific hospitals in the results of treatment is exceedingly promising. Statistics of over three thousand cases, which have been under treatment at different hospitals, indicate nearly forty per cent. restored and temperate after a period of six to eight years from the time of discharge from the hospital. The best authorities unite in considering thirty-five per cent. of all who remain under treatment one year or more as permanently restored. In view of the chronic character of these cases, and the imperfect means of treatment, these statistics are encouraging, and indicate great possibilities in the future from a better knowledge and control of these cases.

Legal Control of Inebriates.—The legal control of the inebriates in America, and legislation, are very imperfect. In Connecticut the best laws are in force, giving power over inebriates to voluntarily commit themselves, or be committed by their friends, without the formality of appearing before a judge or court. In other States they are committed to asylums in about the same way as the insane are. In the hospitals they are controlled legally the same as the insane, only with more difficulty, and the constant intrusion of disputed questions of authority which cannot be settled. Nearly all the leading hospitals have special powers of control, which they exercise with caution in most cases. But generally both legislation and legal authority are not far ahead of public sentiment, and hospital managers are unwilling to go beyond this. Most of the hospitals have power to control patients a certain specified time, agreed upon when admitted to the hospital.

But apart from the view that special retreats are necessary for the sake of inebriate patients, I would urge their establishment in order to relieve the families of such

patients of the burden of maintaining them. I know of
many instances where the wife could provide for herself
and children, but an attack of inebriety in the husband
throws all back. He must be watched day and night while
the attack lasts, medical attendance and medicines must be
procured, the shop or the sewing which would otherwise
bring in enough to keep them is necessarily neglected, and
she is obliged again and again to appeal to friends, who
begin to get weary of giving assistance. The poor wife
says : "Something must be done ; he must be put somewhere ;" but the prospect of prison or of the lunatic asylum
deters her from taking any action, and she says : "Well, I
will give him another trial." These poor women have bitter
lives, and many of them make noble sacrifices for the sake
of their families. The State ought to afford them a ready
means of relief from the consequences of a physical and
mental disease in their husbands for which they are in no
way responsible.

I would then propose the establishment of hospitals for
the care and cure of inebriates, such hospitals to be either
supported by the State, or else under State inspection and
control. That persons suffering from inebriate disease
should, if the outbreak of the disease be recent, on the rep-
resentation of their friends to a magistrate, and after due
examination by medical men and proper certificates being
signed by them, be committed to one of the hospitals for
nervous diseases for a year. That the period of detention
in such hospital might be lessened on the recommendation
of the medical officers. That during the patient's residence
in such institution he should be obliged to do sufficient
work to pay the expenses of his living, and the surplus to be
handed to his friends for his support on his discharge. The
object of the work would not be merely to pay expenses,
but to give healthy mental and bodily occupation ; to substitute healthy nerve-work for unhealthy impulses, and thus
to act curatively. In cases where such seclusion and abstin-

ence from alcohol for a year had been tried without effect
—that is, in cases of presumed incorrigible drink-mania—
the patient should, on each outbreak, if not under proper
care and control, be admitted for treatment until the attack
should be over, and he should be detained as long as the
medical officers thought fit with regard to his safety.

I have explained what proper care and control means,
but will do so again. It means such care as will prevent
the patient getting alcohol during the attack—an exceedingly difficult measure to carry out in private houses. I
would avoid the use of the word "asylum," and name
an institution for the treatment of inebriates "Hospital for
Disease of the Nervous System." It could easily by specified in the rules of the hospital for which particular nervous
disease it was intended. Such a name would be less repugnant to the patient than the word "asylum," and would
serve to educate public opinion. The establishment of such
State hospitals might be objected to on the score of cost,
but it must be evident that the community pays far more
heavily by the present system, or want of system, than it
would do if special hospitals were established. We should
then be saved the expense of maintaining such patients in
prison ; and such special hospitals could be made self-supporting to a very great extent, rendering the estates of those
who were able to pay liable for their maintenance, while the
others should work for their support. Inebriates who might
be allowed to live out of the retreat on parole, could, if they
broke their parole, be treated in the nearest public hospital
and discharged when the fit was over; but if they again
broke out they could be committed to the retreat for the
full time on the recommendation of the medical officers,
without option of leaving until their time should have
expired.

Dr. Day, of Washingtonian Home, makes the following
clear and positive statement of results. Twenty-two years
of experience in this work has demonstrated that the task

is neither hopeless nor thankless ; nor would it be if the measure of success had been lessened one-half from the known rate of percentage of cures. I know of no hospital, infirmary, or reformatory institution in the country—and I may as well say, in the world—in which the proportion of cures to the number of patients treated is greater than it has been annually in this Institution, from the date of its origin down to the present time.

A variety of circumstances beyond our control prevents us from making a fair and just exhibit of our success to the world. A man will go to the hospital or infirmary with a disease or a broken limb, and when healed, or made whole, there is no hesitancy or delicacy on his part, or with his friends and relatives, in making the fact generally known ; sometimes, especially when the patient is of sufficient consequence, the community will acquire this information from the public prints. But it is different with the recovered inebriate, who, for his own or family reasons, shrinks from a confession of his case, and while he rejoices, and is thankful and grateful for his cure, has a natural repugnance to acknowledge that he actually needed the treatment he received. Hence it is, that numbers of men are reformed at the " Home " who return to their families and to their business, giving joy to friends and neighbors, not one in ten of whom know how or where the sudden change was wrought.

As a rule, failures occur with persons who are willing to be cured if it can be done without making any sacrifice or concession themselves towards effecting such a consummation. However honest their desires may be to live soberly, they have not strength of will sufficient to resist strong and repeated temptation. Is not this the case with transgressors of every physical, moral, and Christian law ? . . .

But even such cases are not without hope. Proper medical and dietary treatment may do much to build and strengthen their physical system, but what they need most is medicine for a diseased and weakened mind, and that is a curative agent not to be found in a drug store or a physician's medical case.

In most cases a longer lapse of time is required to effect a cure than the patient thinks he can spare from his business, or afford to pay for. The deaf, the lame, the blind, and the sick, who satisfy themselves with one or two visits from the physician, when their several cases require a patient and lengthened treatment, have just the same right to declaim against the efficacy of medicine and surgery, as the public have to charge us with failure in cases where we have been permitted to treat only for days, when we should have had weeks, and weeks when months were fairly required.

CHAPTER XXIII.

HYGIENIC AND HOT AIR TREATMENT OF INEBRIETY.

There are two classes of drunkards to be found in the United States—one class are drunkards because of their training and education ; the other class is born with a tendency to drunkenness. Let me call attention first to begotten drunkards. These are peculiar. At the time of their conception the father and mother were habituated either in their foods or drinks to the use of some form of stimulants ; not necessarily were these alcoholic or fluid in their nature. Stimulants are substances which contain in their constituent elements properties that when taken into the human stomach by eating or drinking have a specific effect on the nervous system, exciting the heart to undue action, and through it the whole circulatory system, insomuch that the person thus affected takes on an unnatural condition of the vital organs, by which he is subjected to extraordinary activity of such organs.

This condition, continued long enough, becomes habitual, and under it no person can go into the procreative act without carrying over to the offspring a constitutional diathesis or habit of body, or, in other words, a tendency or predisposition to the use of stimulants.

The child thus made up has in him a need for stimulation in order that the vital organs may perform their work to the best advantage, and there will be failure in their action unless this need be supplied. The child may linger along, being feeble ; but, if his habits are such as are common to the children of our country, he will die unless help is given him through stimulation. A physician, if

227

called to such a child, would almost surely discern the need of extrinsic aid being furnished, in order that the vital organs might perform their functions with sufficient directness and vigor to result in good health and continued life.

Under such circumstances, how is a person to be kept from becoming a drunkard? That is a problem which science and morality, separate and together, have been unable to solve hitherto, and will be unable so long as persons permit themselves, while under habitual stimulation, to beget and give birth to children. It is not at all necessary, in order that a child's organization shall have such idiosyncrasy as to call for and require stimulants, that either of the parents shall be drinkers of alcoholic liquors. It is enough that through food eaten or medicines taken, stimulants are constantly introduced into their circulation. Nature, in her great organic processes, knows nothing of, nor cares anything about, the particular substances a person uses whereby his whole system is extraordinarily excited. It answers all evil purpose that this excitement of the nervous system is created and constantly kept up. A married pair can give birth to children in whom shall inhere a tendency to demand stimulation to that degree that a clamor for it is set up in the very center of their bodies, just as surely by the eating of stimulating food as by the drinking of stimulating drinks.

Of begotten drunkards the numbers in this country are larger by far than are the numbers of trained and educated drunkards. There are but very few children now-a-days who have not in them a natural desire for stimulants. Denied these in every form, they put forth but feeble growth, and in many ways show inefficiency in the performance of those bodily functions which must be fully and fairly exercised in order to the production of good health.

Here, then, is an evil of great magnitude which is not touched at all by any temperance movement yet inaugurated, and herein lies the secret of the imperfect success of all such movements. Little or nothing is gained as respects the arrest and overthrow of intemperance while the children who come into the world have not only a natural liking for liquor, but have also what may be called a natural need for it. Every generation repeats the story of its predecessor. As the child grows up his liking for stimulants takes on the form of a need for them. When puberty is reached, the boy, whose activities are then drawn upon more largely and decidedly than at any previous period, finds that to answer to the simple nutrient wants of his system, so far as muscularity is concerned, does not meet his necessities. He may be in good muscle and so of good size of body, but there is a defect in him not to be overcome by force of will when he is called upon to show large, well-directed, and successful energy. He lacks just at this point the agency by which energy alone can be developed, and if he cannot have aid he fails.

It is this cry in his body for stimulation, for something to innervate him and make him feel strong, that sets him to drinking ; and this clamor is just as likely to be manifested if his father and mother are total abstainers from all intoxicating drinks as though they indulged habitually in them. Total abstainers from these drinks, who do not abstain from other stimulants, may use them to a degree and in a measure that will produce quite as deleterious effects on the nervous system as would be produced by drinking alcoholic beverages. So the child born of parents who are pledged to the entire disuse of intoxicating liquors, may have a natural desire or need for alcoholic stimulants, as truly as the child born of parents whose nervous systems have been habitually excited by alcohol.

The object then to be sought must be, not the exchange of stimulants, but the abandonment of them altogether. Then children will be brought to birth with no tendency in them to the use of stimulants, and then they can be trained up and reared along a line of sobriety and abstinence from

all stimulants, which will make it quite out of the question
that they should ever use intoxicants in any form.

There is a process by which a child with an inborn need
for stimulants, in order to make his organs perform their
natural functions, can be reconstitutionalized. That process involves freedom from taxation of the nervous system
for years after his birth. The child has to be cared for and
looked after like a young animal. The developments that
need to be made in him are such as affect the nutritive
nervous system. He should be kept free from all cerebral
excitement. He needs to be cared for by judicious and
wise nurses. He cannot be permitted to eat or drink, nor
do or be, as children usually are.

If he were thus related to life he would die ; but if he
can be kept free from all unnecessary excitement of brain
and be handled with close watchfulness for the first three
or four years of his life, his relations to vital development
will become greatly changed, and from that point on, the
reconstructive processes, so well begun, will proceed to completeness. He will ultimately reach a condition of his
nervous organization where he will no longer need stimulation.

Till that is accomplished, attention must be given to him
or, with no stimulants administered in his food or drink, he
will fade away and die. It is most unfortunate that a child
should be made up in a state of dependency upon factitious
aids, but where such dependency exists, reconstructive measures can be instituted and made effectual. This I have
proved on this Hillside to the satisfaction and joy of a great
many parents. The better way, however, is to forestall such
an issue as this, and not have children begotten when the
parents are under a bodily habitude of subjection to stimulation.

For a long time I have been convinced that the present
temperance reformation is radically inefficient. The unphilo-
sophical aspect of it will have to give way before a broader

and more effective effort, or we shall have to keep it up forever, having in each succeeding generation as many or more persons becoming drunkards as in generations preceding.

The second class of drunkards—those who are made so by education and training—are persons not born with a constitutional necessity for the use of stimulants. As children they do not in our country constitute a very large class. They never need be drunkards. The temptation to drink is not in their bodily constitution. It lies in society in some or other of its forms of associations. Thrown into good and upright relations from childhood, they will not become drunkards. They will not even drink liquors moderately. As they develop into intelligence and moral sense, they will see that the use of stimulants for purposes of exhilaration, or of innervation of the nervous system, is not good for them. They will be ready to join the total abstinence ranks, and when they have joined them, they will stand true.

But out of this class there is a certain number who do become drunkards. These lack the early training and right association and proper surroundings. They eat highly stimulated foods and drink stimulo-narcotic drinks, like tea and coffee, until they have created in them an abnormal desire for things that are exciting. No person can habitually eat largely of the indiscriminate foods used in most families, and drink such beverages as are there daily imbibed, without after awhile coming to have a longing for stimulation. Young persons come to have a want for something stronger than tea or coffee, and many of them have opportunity at the home table to indulge in the milder forms of alcoholic drinks. So they go on from bad to worse and from worse to worst. Out of this class a certain proportion become drunkards and once they are habitually inebriated, their restoration to permanent sobriety is as difficult as though they belonged to the first class.

This cure does not lie with certainty in any movement

which simply makes its appeal to the moral sense. A
drunkard may be said to have no moral sense. He has a
physical sense and he has a social sense. His physical sense
tends directly to keep him a drunkard. His social sense
may work for his recovery. He may be so situated that
social influences shall operate to induce him to sign a pledge
that he will not drink any more, and here and there may be
found one out of a great many who will keep such a pledge,
not by reason of any inherent strength which he himself has
whereby to keep it, but by reason of the outside influences
which affect him and the watch-care which is had over him.

If thinking men and women will take pains to investigate
this matter as thoroughly as I have done, they will see that
a very large proportion of the whole number of drunkards in
this country is made out of the class which is born with a
liking for, with a tendency to, and with a need of, the use of
stimulants.

It is now coming to be recognized that inebriety is a
disease, but its treatment has heretofore been too much
relegated to the moralist instead of the physician to whom
it properly belongs. The moralist has given us theories as
to its cause and cure, but the theories have brought us no
advancement ; the disease still remains to vex their patient
souls. The medical profession, however, place this subject
in its true light, and give a hope to the world that in a short
time there will be a better way to handle this whole matter.
Dr. Wright most forcibly says : "Drunkenness is in every
essential particular a condition of civil death, and it would
seem best that it should be so construed by the law of the
land." Society should be protected from the habitual
drunkard, and especially should he be protected from himself. To my mind the bane of the age is excessive alimen-
tation, leading up to a desire for stimulants, which is
naught but a morbid craving, and in ignorant hands stimulation is supposed to ward off its consequences.

On the contrary, and in reality, the use of stimulants at

such times induces a condition of internal inflammation that increases the desire for further stimulation, and also the inability to properly dispose of whatever aliment may be present, thus effectually rendering a bad matter worse. We well know that nothing will so speedily subdue the nervous storm as the ever convenient and alluring alcohol. It is sought to give insensibility to nerve agony, and secures for a time, rest and repose, but the wear and tear of this oft repeated nerve strain is frequently shown in paralysis and sometimes insanity. Perhaps the most serious effect of alcohol is its direct tendency to interfere with nutrition, and by promoting growth of cellular tissue to compromise the integrity of the brain tissue where the poison is not readily thrown off, and where it soon destroys not only its co-ordinating power, but degenerates the brain substance. Inasmuch as the mental and moral character of the individual depends upon the action of the brain itself, we cannot look for good results when there is any impairment of its substance ; on the contrary, we often see the worst results from such a condition.

The impairment of consciousness is only one of the many forms by which the influence of alcohol is felt, and suggests the question whether we have any process whereby its elimination may be successfully secured. We also know that with the inebriate there is a lack of fine moral sense, not infrequently amounting to obtuseness, and that this condition is far reaching, affecting even the progeny, thus making it hereditary. If there is a process whereby the blood itself can be purified, we may, with every reason, expect the brain tissues to participate in the advantages derived therefrom, and consequently we will have greater clearness of perception, followed by a quickened moral sense. The irresistible impulse of our modern civilization, from infancy to old age, is push, and the mental and physical powers alike suffer in the long run. The free use of alcohol is accountable for a large measure of this condition. Can we not teach the people to give more time to rest and recuperation and less to stimulation? If it is wished to place the inebriate in the condition most favorable for cure, it is important that there should be institutions created for that purpose.

Granting that inebriety is a disease, our efforts should be to eradicate that disease, and in order to do so, there must necessarily be desirable surroundings as well as control over the patient. The model institution is yet to be built, wherein the hot-air bath shall hold a pre-eminent position, where narcotics shall be entirely disallowed, even though it may be an improvement on inebriism to have one's system saturated and senses blinded by narcotism, and when there shall be enough control to prevent any dallying with the tempters. The theory of the action of the hot-air bath is very simple. Like the action of the sun's rays upon Bunker Hill Monument when shining upon one side and causing it to lean toward the other, so does this agent act gently and yet powerfully. The primary action of heat, which is the one essential thing of the Turkish bath, is to relax the tissues of the body and thus invite a more perfect circulation to every part of the system; by a more active circulation every sense is quickened—the secretions are more thorough, the excretions more perfect, the blood is better supplied with oxygen—the skin assumes its natural roseate complexion, indicative of the improved condition, and each and every function, whether it be that of the lungs liver, spleen, or bowels, comes in for its share of the general benefit—in a word, it opens every pore of the skin, and hence, the most perfect sewage to the body. The secondary action is that of profuse sweating—where water from the blood and debris or used-up tissue and poison held in solution are rapidly thrown out of the body.

According to recorded observation, "the quantity of blood in the body is lessened by the free excretion which takes place through the skin and lungs; the body weight is reduced, and the work of the heart in this way lightened, at the same time that its substance is better nourished by the improved quality of the blood supplied to it. The peripheral arterioles of the body, too, become dilated and filled with blood, thus affecting a corresponding emptying of the blood vessels of the internal organs. Lastly, as a result of the alternate warm and cold douching, the vaso motor energy of the vessels is increased, thus rendering them more capable of resisting any strain thrown upon them." Thus it will readily be seen how quickly congestion, wherever it may be located, is broken up and the offending material thrown out through the pores of the skin. Under such conditions absorption and elimination have their most perfect opportunity and equalization crowns the work. It must be apparent that alcohol is soon eradicated from the system under such favorable conditions, and that torpidity gives place to activity. Furthermore, no living tissue or vitality can be abstracted by the process—nothing is thrown off but what the system is better without. One bath has been frequently known to relieve an intermittent pulse, giving a smooth, regular action to the heart, indicating a well balanced circulation. What known drug can do this in the space of half an hour? And another great advantage in favor of this treatment is, that there is no poison left in the system to work its way out, as is sometimes the case when drugs are administered; per contra, the individual is left in a calm and quiet frame of body, which necessarily reacts upon the mind. It could not be maintained that the hot-air bath would renew brain or other tissue where there has been actual lesion, but it will place under most favorable condition for repair what is left, and then adjacent or collateral parts will do their best to carry on the work of the injured part.

By placing the patient in an institution of the kind mentioned, we at once completely remove the cause of the disease, and then with the bath we have only the effects to treat. In the instance of insanity the hot-air bath has in a large number of cases brought relief to deranged conditions and given harmony to disturbed mental functions, and this where the cause was present, for we know that this disease obliterates the patient intellectually, and leaves the physician in the dark in reference to the cause or the effect of the malady. It must, therefore, be evident to every medical mind, that the remedy which will effect every organ and create in it an action to throw off diseased conditions, is the only one to meet such cases. With how much more reason must we expect even better results in cases of inebriety, where the cause of the disease is eliminated and the advantages of treatment are more perfect. The mucous surfaces of the inebriate, and, in a minor degree, those of the moderate drinker, are in a chronic state of inflammation. The effect of the hot-air treatment is to reduce that inflammation by purifying the blood, thereby relieving that immoderate craving for stimulants, that only perpetuates and increases the disease instead of giving relief. During my long experience in the administration of the Turkish bath, many persons who had been more or less under the influence of alcohol, have expressed to me in most unqualified terms the benefits that they had derived from its use, particularly those who came after a debauch. In fact, this has been thoroughly demonstrated, as the experience of all bath establishments will testify. Place man or boy in a clean suit of clothes throughout, and he will not only take good care of the clothes, but also of himself, and behave more discreetly than before.

In like manner, if you thoroughly purify a man, as is done by one of these baths, he at once realizes that he is a cleaner man and on a higher plane, his senses are more acute, he is in his best condition, he respects himself so much the more, and is less liable to return to his base practices. It is stated as a fact that in no country has inebriety been found co-existent with the bath. Temperance and cleanliness are its handmaids. This treatment has had but a limited trial in this country, though it has been successfully used in a multitude of cases in Great Britain, particularly at Dr. Barter's establishment, near Cork, in Ireland. The only demonstration in this country was at Binghamton, during the first three years of it administration, in which time not a death occurred among the patients. Dr. Lees, says: "In the case of persons having latent cravings for drink, we know of few things more efficacious than a short course of that peculiar method of cleansing, which, borrowed from the Orientals, has been recently introduced into many cities—we mean the Turkish bath. Who, suffering from morbid accumulations incident to town life, that has ever tried these processes, has not felt a wonderful increase in the vital elasticity of his frame ! It is as though a heavy weight had been lifted from the bent spring of life, permitting fuller and freer play to the vital machinery and creating a feeling of sympathetic purity in the soul."

The true physician stands before the community in the light of a teacher as well as a healer, and his opportunities for usefulness are large and often far reaching. Probably no class do more charitable work than the men of this profession. In no way can they do more good than in encouraging both by example and precept, those institutions that have for their object the welfare of the community, and no institution of modern times promises so much to the mass of the people as the genuine hot-air bath. Sanitary science is of incalculable value to each and every one of the community, for it deals with that which is vital to the well-being of the whole, but the hot-air bath is sanitary science, refined and brought to the individual ; indeed it is the perfection of sanitary science. As a prophylactic, it stands at the head of all remedies. As a disinfectant, none with it can compare. The more it is popularized, the nearer it will come within the reach of everybody, and the more widespread, necessarily, will be its blessings.

CHAPTER XXIV.

DUTY OF THE STATE IN THE CARE AND TREATMENT OF INEBRIATES.

The State assumes the right to license reputable men to sell alcoholic liquors to those who are in the habit of using them, to those who would like to experience their effects, and to those who wish to treat their friends to something uncommonly good. This the State does in accord with the laws made to suit public opinion.

If this "opinion of the public" would demand it, the State could very properly assume the right to license opium dens, or saloons of euthanasia, where a fellow could shuffle off this mortal coil according to the most approved methods.

In other words, public opinion forms laws, whether just or unjust, and it is the duty of all good citizens to acquiesce in a state of affairs brought about by the decision of the majority.

In a study of this topic I enquire, *First*, Has a man the right to become a drunkard, and after he is a drunkard has he the right to be one?

The laws of the State permit the drunkard to be made, because the State licenses reputable men to sell alcoholic liquors ; and the laws of the State concede the right to become a drunkard to every citizen, because the State says that personal liberty should not be interfered with. On the other hand, the laws of the State direct that the drunkard be fined and imprisoned for being what his personal liberty entitles him to be. It is, therefore, logical to say, that

239

according to the laws of the State, a man has a perfect right to become a drunkard, but that after he is a drunkard he has no right at all to be one.

Science claims that the inebriate does not exercise his free will to remain or to be a drunkard, for the simple reason that he no longer has a free will to exercise, but that he is the involuntary slave of an uncontrollable desire. Science also claims that the drunkard may have never exercised his free will in the matter. He may not be a drunkard from choice, but he may have inherited a predisposition to become a drunkard, and necessity, opportunity, and circumstances may have made him what he is, an habitual drunkard.

Science further claims that it has demonstrated that inebriety is a disease ; that the State licenses the making of this disease, and that the State does not judiciously recognize inebriety as a disease.

Science further claims that no man has a right to become a drunkard, because, as a rule, every man has a choice in the matter, and he ought to choose what is best for the individual and for society.

Science further claims that " Punishment is no cure for the Disease of Inebriety."

The State and science, therefore, differ in their ideas about "becoming" and " being" a drunkard. The State considers "becoming a drunkard " a personal right, and "being a drunkard " a crime. Science holds "becoming a drunkard " to be a sin, and "being a drunkard " a disease.

Second. Is drunkenness a disease ?

The laws of the State sanction moderate and temperate drinking. The men who made the laws did it in accord with the wish of the public. And the public, no doubt, is satisfied that the State cannot legislate its people into temperate habits. The people must be educated to be temperate through the press, the pulpit, the school, and the lecture.

There are men who are said to get drunk by accident. These are indiscreet and should not be judged harshly. They ought to beware of accidents lest they become habitual drunkards.

The habitual drunkard, however, suffers from a disease called dipsomania.

Habitual drunkenness or dipsomania may be inherited or it may be acquired.

"Dipsomania is a mental alienation due to a morbid condition of the nervous structures, generally hereditary. The strictly periodical form of this type of dipsomania, the tendency to gradually shorten the intervals as the years pass, and the peculiar mental condition preceding the debauch, are a proof that dipsomania is a disease of the cerebral nervous centers analogous to recurring neurosis, such as epilepsy, etc.

This disease is nothing but an attack of uncontrollable drunkenness, always kept up until the stomach refuses longer to tolerate the alcoholic drinks. Then the attack stops as suddenly as it came, the sufferer recovers his usual health and spirits and enters into his business in a way as if nothing had happened. As a general thing these attacks recur at intervals of from one to six months, and the end is, some disease of the renal, hepatic, or gastric organ carries off the patient.

Earnest resolutions or pledges do no good to ward off the attack. When the time comes the patient succumbs. An indescribable feeling of weakness of the nervous system is generally the first sign of an attack. This may be brought on by over-work, over-study, anxiety, worry, trouble, anger, etc., and the patient thinking himself proof against a debauch by his long interval of sobriety, yields to the temptation, and then nothing can head him off. Friends, family, duty, rank, morality, resolution, and pledge are all forgotten, and the patient will drink as long as his stomach will bear it. So strong is this desire for drink, while the attacks last, that the patient will drink as long as he has money, or rather as long as he can get the liquor, though he may have to beg, borrow, or steal it.

Such people are the despair of their friends, the torment or ruin of their families, the scandal of their community. Seventy times seven they fall, and are lovingly raised up. They express contrition, they make firm promises of amendment, but they always fall.

From the researches of many authors it appears that inebriety is a nervous disease closely allied to insanity, which manifests itself either periodically or constantly. It may commence suddenly as a consequence of some severe shock to the brain. The disease may also have its origin in the social habits of the patient, who from a simple convivial drunkard may become transformed into a regular inebriate. It may be produced by the action of other poisons besides alcohol, so that there may be as many varieties of inebriety as of narcotics. There must, however, be a predisposition to inebriety in order to effect its evolution. Healthy men without neurotic predispositions may drink voluntarily in moderation without thus breaking down ; but an inter-current disease may turn the tide even against such individuals, and if they do not themselves suffer the penalty of indulgence, their children will be found far on the road that leads to inebriety. Hereditary influences are among the most potent that determine this disease and they follow the usual course. Thus, in mixed families, the male children of an inebriate mother, or the female children of an inebriate father, may alone exhibit the morbid tendency.

There is another type of drunkenness which goes under the name of ebriosity, and by which is understood the condition of continual half-way intoxication. This, necessarily, occurs only among saloon-keepers and those engaged in the liquor traffic, having access at all times to alcholic drinks. This is a very fatal form of drunkenness, and may be called incurable. It is the form of drinking which life insurance companies especially fear.

Habitual drunkenness is, therefore, a disease brought on by the excessive use of alcoholic drinks, though this use may or may not have been continual, and the victim is an involuntary slave of an insane propensity. He knows what is right, but cannot choose it ; and he knows what is wrong, but cannot shun it. There is no loss of the power to judge of right and wrong nor any disturbance as to facts, but the mind is powerless to control conduct according to knowledge. This state which the drunkard is in may be called criminal irresponsibility.

Third. Can the drunkard cure himself or be cured under the existing circumstances ?

In the early stages of dipsomania, inherited or acquired, something can be done for the patient, but the cures are the exception.

As to chronic cases of drunkenness a reformation is improbable and very nearly impossible under the existing circumstances. The temptations are too great and the opportunities are too many, and though the drunkard may have a desire to reform or to cure himself, his will-power is enfeebled and he cannot resist the demands of his habits unless he be removed from temptation and it be made absolutely impossible for him to get a drink. The drunkard is an object of contempt and disgust as a drunkard, and an object of pity and danger as a man, as a father, as a son, and as a brother. He is pitied by all good men, but they are powerless to help him. He cannot help himself.

He cannot reform while he is in the midst of temptation. The licensed liquor houses are easy of access, and indiscreet friends are not wanting to tempt him to go into such places and to take but one drink, which is the spark that lights the attack.

Dr. Kerr says, " The struggle of the intemperate for freedom is a combat more terrible than any other fight on earth." It is a hopeless fight, I add, if unassisted.

Dr. Parrish says, "The temptation with which they are tempted is within. It is subjective. It circles in the stream that gives them life. It may be likened to a battery that is hidden somewhere in the cerebral substance—connected by continuous fiery wires, with a coil in every ganglion, from whence they continue to extend—attenuating and distributing as they go, reaching after the minutest nerve fibrils, which need only a throb from the inborn impulse to transmit a force that quivers in every muscle and burns in every nerve till the victim is suddenly driven to debauchery."

Tell me now, is not the condition of the drunkard a deplorable one? In fact, is it not a blot on civilization? Should not something be done to prevent the increase of drunkenness and to diminish what there is of it?

Fourth. If the drunkard cannot cure himself, has he a right to be protected against himself and against those who are licensed by the State to sell him the wherewithal to remain a drunkard?

It is true that there are a great many private institutions throughout the country for the cure of inebriates, and they do a great deal of good, but the charges for treatment in these institutions cannot be met by the majority of drunkards. Taking into consideration that less than a year's treatment will do no good, it is quite plain that the majority of these unfortunates have, under the circumstances, no chance whatever offered to them to become cured of their malady. This is a serious matter to those afflicted with the drinking habit, and it is in their behalf that I make this appeal to the citizens of Indiana, relying on their feelings of justice, charity, mercy, and humanity that my efforts in behalf of the habitual drunkard will not be without success. I ask for these unfortunates nothing but what is just. There is no hope for these poor creatures until they are withdrawn from temptation and placed under restraint.

Fifth. After a man becomes a drunkard is he a dangerous man to society? and, if so, has society a right to be
protected against the drunkard?

The State protects society against the drunkard in the
following manner:

1. The law assumes that he who, while sane, puts
himself voluntarily into a condition in which he knows he
cannot control his actions, must take the consequence of
his acts, and that his intentions may be inferred.

2. That he who thus voluntarily places himself in such
a position, and is sufficiently sane to conceive the perpetration of the crime, must be assumed to have contemplated
its perpetration.

3. That as malice in most cases must be shown or
established to complete the evidence of crime, it may be
inferred from the nature of the act, how done, the provocation or its absence, and all the circumstances of the case.

4. The law has not yet judicially recognized inebriety
as a disease.

The State does in every way try to prevent the making
of robbers, thieves, burglars, and murderers, and criminals
in general, but it licenses men to make the drunkard. The
State protects society against the murderer and the robber
by imprisoning him, but it allows the drunkard to constantly menace the well-being of society.

Would you say that the poor wife, who supports herself
and her children by sewing and washing, has no right to be
protected against the evil habits of her husband? Would
you say that the drunkard's children have no right to be
protected against the evil influence, the bad example, and
the burden of drunkenness of their father, which threatens
to blight their whole lives? Would you say that the friends
of the drunkard have no right to have an asylum provided,
wherein they may place their unfortunate son or brother or
sister or father or mother or friend, in order that they may

be cured of their malady, which is a curse to all coming in daily contact with the habitual drunkard?

These are questions that can have but one answer. Yes; society should be protected against the drunkard, who is made by consent of the State in accord with public opinion.

Sixth. If society has a right to be protected against the drunkard, if the drunkard has a right to be protected against himself, and if the drunkard has a right to be protected against those who are licensed by the State to sell him the wherewithal to remain a drunkard, does it not follow that the State should afford the protection?

In answer to this question I read you several extracts:

" From the very nature of the malady it is scarcely to be expected that the inveterate drunkard will voluntarily submit to control, or continue under it for a sufficient length of time to receive lasting benefit; and therefore it seems essential, as in the case of other insanities, that legal power, with, indeed, the neglect of law to provide such a check and remedy, seems inconsistent, unjust, and inhumane when we consider that while it permits the insensate drunkard to endanger his life, to waste his property, and deprive his family of that which they are justly entitled to expect from his hands during life, or to fall to them at his death, it holds him responsible for any criminal act he may commit. No doubt the law assumes that he drinks voluntarily, and with his eyes open to all the consequences, and that his practices therefore form an aggravation of his guilt; but such is not the case, for he drinks involuntarily, and he is unable to exercise his reason aright or govern his will."

" All experience has shown that little progress or none can be made toward the permanent recovery of a dipsomaniac so long as his business places him in more or less contact with alcoholic drinks, or in frequent association with drinking comrades. Consequently, both physician and friends should combine their influence to separate as far as possible the patient from such associations. And if it cannot be done in any other way, let him be induced to take a residence for six or twelve months in a well-regulated asylum for inebriates until the paroxysmal tendencies have been broken. Enforced seclusion in a proper asylum, with no possibility of obtaining any kind of alcoholic drink, but where good air, good food, kind treatment, and some suitable employment can be furnished, on the same principle that applies to the treatment of insane persons, will save them from early destruction."

"For the permament cure of inebriety, however, nothing avails but special treatment in hospitals provided for this class of patients. Of these the number is increasing as the public becomes informed regarding the nature of the disease and the appropriate means of combating its ravages."

Dr. Carpenter says : " However responsible he may have been for bringing the disease on himself, his responsibility ceases as soon as he comes under the influence of the malady. The disease, however, may not be brought on by the act of the individual, and then it is clear at once that neither directly nor indirectly can he be deemed responsible. But, suppose that it were the result of his previous conduct, I repeat that, however culpable he may have been for that, he is not a responsible being while afflicted with the malady ; for I can see no distinction between this form of the disease and any other which has been induced by the habits or acts of the individual.

" The only chance of a cure or alleviation is from attention to the health and abstinence from intoxicating liquors. Neither can he be cured so long as the patient is at large, and no amendment can be depended on, unless he has undergone a long course of discipline and probation. Considering, then, that the individual is irresponsible and dangerous to himself and others, and that his disease can be treated only in an asylum, it is not only merciful to him and to his relatives, but necessary for the security of the public, that he be deprived of the liberty which he abuses and perverts, and that he should be prevented from committing crimes instead of being punished—or, I should rather say, being the object of vindictive infliction—after he has perpetrated them.

"Of the chronic form I have seen only one case completely cured, and that after a seclusion of two years' duration. In general, it is not cured. Paradoxical though the statement may appear to be, such individuals are sane only when confined in an asylum."

The insane asylums cannot and ought not to be used as a home for inebriates. Only those in the last stages of alcoholism, that is, only those who are actually insane, are sent there. Special attention cannot be paid to cases of inebriety in the insane asylums ; besides, no one would voluntarily apply for admission into an insane asylum, and if he were to do so, it were questionable whether he would be admitted. Therefore, the insane asylums offer no aid to the inebriate.

Seventh. If the State should afford the drunkard a chance to be cured of his malady, how can it be done ?

The only way in which the State, under the existing circumstances, can provide a protection for the drunkard, so that he may be cured of his malady, is by erecting and maintaining homes for inebriates. The sending of confirmed drunkards to these homes should be made compulsory by laws in the same manner as the insane are sent to the asylums. Voluntary entering into the homes by confirmed drunkards should be encouraged, but only allowed after a proper examination by two physicians, who should certify as to the condition of the patient. A complete history of the case would also help the medical superintendent and his medical assistants in treating the patient.

The inebriate must remain at the home at least a year, when the medical superintendent, under the advisement of the board of trustees, may let the patient out on trial until he has proven himself able to govern his will. The friends should not be allowed any control over the patient after he becomes an inmate of the home to which, according to circumstances, he is assigned.

It is not my object to direct just how these homes should be built and afterwards carried on ; all this is subject to the decision of those who make the laws ; but I feel that in order to make this paper of sufficient worth to merit attention, I must give a general outline of what can be done for the inebriate. With this object in view, I will suggest that the money derived from the licenses issued and from the fines collected from those who are not yet confirmed drunkards, and from those who transgress the law by selling to minors, by selling on Sundays, by selling without a license, etc., be used for erecting a home for inebriates of the male sex, and another for the female sex, and for the sustainment of these homes.

The homes, after they were once in good working order, could be made self-sustaining under efficient management. Indeed, the inmates should be employed in labor, in order to learn discipline and improve their bodily health. They should also be afforded recreation and amusement to prevent discontentment with the situation in which they are placed, and to convince them that there are other ways of enjoying life besides sitting in a saloon and becoming drunk.

These homes should be erected in the country, away from temptation. Proper rules and regulations should govern the inmates as well as those in attendance. The superintendents should be medical men, who ought to be under the control and advisement of a board of directors or trustees, one of whom should be the President of the State Board of Health. The salaries of the officers should be regulated by law, and they ought to give a bond. Politics would necessarily have something to do with the appointing of these men, which, undoubtedly, would be for the best.

The drunkard who is out on trial should be considered as belonging to the institution until he has kept sober for the time of one year. Should he show signs of return of his malady before the year is up, he should immediately be brought back to the home. It is presumed that the person who can keep sober one year is no longer to be considered an habitual drunkard.

CHAPTER XXV.

CARE AND CONTROL OF PAUPER INEBRIATES IN TOWNS AND CITIES.

A certain proportion of the population of towns and cities are composed of what are called the " Criminal Classes "—those that require the constant espionage of the police, and the adjudication of justice. Intermingled with these in no small proportion is the pauper inebriate—friendless, home-less,—appearing in various rôles on the public stage as drunkard, tramp, or vagrant, many times entered on the blot-ter of the police station as an " habitual drunkard," or "rounder,"—appearing at the various hospitals and dispen-saries with disease or injury incident to his habits—finally we find him in the wards of the charity hospital or among the chronic insane of the insane asylum, if perchance sudden death from natural causes or suicide has not intervened—and whether his career terminates on the street or in the hospital, or the cell or asylum, the trench in " Potter's Field " receives him, and thus the story ends.

During the year 1887 the department of police and excise of the city of Brooklyn report 23,912 arrests ; of these arrests 13,862 were for intoxication ; of these, 108 are stated as habitual—we presume this to mean that they were known to the police as " habitual " drunkards, but it will at once be seen that this is entirely out of proportion to the number arrested. It would be no risk to say that of the 13,754 intoxicated persons arrested, many, if not the larger proportion, were habitual or periodical users of alcoholic stimulants, and that to intoxication.

The English testimony as to the relation which alcohol bears to the so-called criminal classes is very conclusive. In the " testimony of chief constables and superintendents of police," taken before " The Committee on Intemperance for the Convocation of York," in 1874, in reply to this question—" What proportion of those who have come under your cognizance as criminals have been the victims of drinking habits and associates."

A. " If by the term criminal is meant persons convicted of any offence against the law, sixty-five or seventy per cent."

B. " Nearly all."

C. " Fully nine-tenths."

D. " Quite nine-tenths."

E. " Twenty per cent. of the summary convictions of one year are absolutely for drunkenness—exclusive of a large proportion of the residue attributable to drunkenness."

F. " Nearly half the entries."

G. " About three-fourths."

H. " During the past twelve months in this division there has been 283 persons apprehended for serious offences. I can safely state that 200 apprehensions were directly caused from the effects of drink."

Question—" What proportion of those taken into custody are under influence of liquor ?"

A. " 25 per cent. in country, 70 per cent. in town."

B. " Those directly arrested and those summoned, all cases, 90 per cent."

C. " 70 out of every 100 persons when arrested are drunk."

D. " 161 persons arrested in this district in one year, 75 were under influence of liquor."

E. " 50 per cent. are apprehended as drunks and disorderly independent of any other offence."

F. " The majority of persons arrested and charged with drunkenness. I should say 70 out of every 100."

We then have the testimony of chief constables, superintendents of police, governors of goals, and chaplains, that at least two-thirds if not three-fourths of all arrests made by the police, the persons were addicted to the use of alcohol, and that a large proportion of these were intoxicated when arrested.

If we were to consult the police and criminal records of any of our large cities, New York, Philadelphia, or Boston, we might not equal but we should certainly approximate such testimony as that given before the " Convocation at York." We cannot then shut our eyes to the fact that in every city and town a certain proportion of the population are more or less continually under the influence of alcohol, and that to a degree often dangerous to the community at large. Intoxication, with or without overt criminal acts, continually occurs, rendering it necessary to arrest and imprison this class.

The question now before us is whether the present method of dealing with the inebriate is the best, and if not, what are its disadvantages. Those who have given thought to the subject confidently assert, that the present method of arrests, fines, and short term imprisonment (or occasional six months) is not the proper and scientific way of dealing with the inebriate. By this method, on regaining his liberty the individual simply repeats his act of intoxication and is again subject to arrest, fine, or imprisonment ; after this has been repeated several times, he is known as the " repeater," or " rounder." Instances are on record where one person was subjected to arrest for intoxication over one hundred times, a period, of course, extending over some years.

A female is reported as having been convicted forty-eight times for various offences, at all times committed through drink. She paid £200, or $1,000, as fines for drunkenness. The large majority in English prisons of re-committals are due to intemperance. If crime be associated with the intoxication, as assault, grand larceny, then the

chronic inebriate, strange to assert, will get the best treatment ; the law will give him the full benefit of his criminal
act. Restraint, and a long-continued period of total abstinence will be enforced during his term of imprisonment,
and when his sentence expires, he will often leave greatly
benefited, and practically a sober man. Instances are on
record where the inebriate has requested that he might be
placed in prison and thus secure restraint, seclusion, and
the discipline of prison-life, and thus attain habits of sobriety.

The testimony from English prisons is singularly unan-
imous on this point as well as conclusive.

Question—" Do you consider the health of patients would
be affected by total abstinence from intoxicating drink ?"
Governors of jails testified " prisoners are universally bene-
fitted ; there are cases where it might be occasionally used,
as in feeble and broken-down prisoners—but these occasions
were rare." As a rule, men who have served long periods
of imprisonment, and who have been habitual drinkers, go
out heavier and better in health than they came in. Their
general improvement in health is due to cleanly habits,
warm bath, good ventilation, regular rest, systematic exercise, connected with prison discipline, and *total abstinence*
from liquor of all kinds.

This result incidently points out clearly, we think, the
general plan and method by which we ought to control and
treat the pauper inebriate. He ought to have all the
advantages that prison discipline may secure to him without the necessity of a criminal proceeding on his part.

Undoubtedly, the law, in dealing with the inebriate,
simply as an inebriate, is faulty or only partial in its effect
upon him. "It practically does this, it arrests him, and
fines or imprisons him for a short period—too short for any
benefit to be derived from it,"—and then lets him go. The
law is like an incompetent physician : it first makes a wrong
diagnosis, and then prescribes an inert and therefore inef-

fectual remedy. It reprimands the inebriate, it does not treat his case at all. It looks upon the inebriate as an individual who has the knowledge of right and wrong, and full power of volition ; it regards the act of inebriation as deliberate and voluntary, and therefore it proceeds by fines and imprisonment to lash back into moral decency and rectitude the offender.

But do we not recognize the value of restraint ? Would we permit the brawling drunkard to make night hideous, or the insane drunkard to scatter ruin right and left ? Certainly not.

What plan would then have all the advantages of the present system of dealing with the inebriate, and none of its disadvantages ?

In the first place, as to *arrest* and *restraint.* The inebriate should be arrested if found intoxicated upon the street or any public place, or upon a warrant issued on due complaint of his family, or in case they failed to do their duty, by a committee of reputable citizens of the ward in which the inebriate was a resident, or the officers of said ward. A warrant should be issued on complaint from any of said parties, by the proper justice, and the inebriate arrested.

Proper testimony should then be secured as to facts concerning his inebriety from reputable medical and other sources. He should then be sent to an *inebriate reformatory, hospital, or work-house,* for the institution should include all these features ; we are dealing with a diseased person, not a criminal, but as a pauper inebriate, without friends, or if he has friends, without means.

There are abundant provisions for the wealthy inebriate. Private asylums are numerous, and the appointments are very complete for his treatment both here and abroad. It is the pauper, his more unfortunate brother, whom we are considering.

The pauper inebriate is now duly arrested ; he must be restrained and controlled for some definite time in some institution. The period should be not less than one year, made longer if necessary by recommittal. The institution to which he is committed should be placed in the suburbs of the city or town, with convenient access to it. Abundant grounds should surround the building, or, better still, a farm should be the site of its location. Out-door occupation, so beneficial in the treatment of the chronic insane, would be no less so in the case of the inebriate. A competent medical superintendent, with suitable assistants, could readily conduct such an institution. Its inmates would be chronic inebriates ; all insane persons, or those incurable from other diseases, should be sent to their proper asylums or hospitals. Such an institution, with properly appointed work-shops, a farm under cultivation, well-stocked and planted, with a practical farmer at the head to regulate the labor of the inmates, would be almost self-sustaining, for the inmates would not be like those in an insane asylum, mentally inefficient, or those in a charity hospital, physically helpless, but many would be skilled workmen who outside would command good wages. Then, also, a system of payment for extra work might be made, so that when they left the reformatory something would be due them, and they would not be turned out paupers. If this plan were adopted, a large body of chronic inebriates that now drift about in the community would not only be restrained but made to a certain extent self-supporting, and in a certain proportion of cases cured.

The inebriate Home at Fort Hamilton is based on some such plan, and demonstrates on a moderate scale what might be accomplished on a larger one.

Every large town and city should have such an institution of sufficient size to meet its wants, containing a farm, a work-house, and suitable medical care. It should be readily accessible, although the price of land would regulate somewhat the site of its location.

The locality should be healthy, and the internal and sur-

rounding sanitary conditions good.   The dietary should be
generous, of good quality, and the food well-cooked.   This
is essential.   Specialists in lunacy have found that a certain
way to precipitate acute or sub-acute lunacy into the chronic
forms is to put the patient on low and innutritious diet.
Out-door exercise and occupation, as well as those measures
that will eventually appeal to his better nature, lead him
back to thoughts of home and family, develop his higher
tendencies, prompt his aspirations, and raise him above the
mere animal life he has led so long.   To deprive such a one
of religious privileges, or the intellectual enjoyment he may
crave, is a refined species of cruelty that no true form of
philanthropy would be guilty of or tolerate.

This is no sentiment, it is practical fact and truth, for
among these pauper inebriates are found lawyers, editors,
physicians, clergymen, writers, artists, and skilled artizans,
men who have fallen from high estate.   It is natural, then,
with returning and improved physical and moral perception
they should begin to crave that which feeds the intellect
and administers to the improved moral tone.

So much, then, for the Reformatory, the work-house, the
hospital, where we would place the chronic inebriate without friends and without means.

But while the above institution will care for the chronic
inebriate, it does not, and cannot, fill a want severely felt and
long needed—how shall we deal with intoxicated persons
arrested on the streets by the police?

The usual method is to arrest them, take them to the
nearest police station, prefer a charge of intoxication, with or
without disorderly conduct, record the case on the blotter,
and commit the accused to a cell, to await the sentence of
the justice.   The following morning he is brought before
the police court—if a first offence, and not particularly ag-
gravating, sentence may be suspended—usually a fine is inflicted—and if this cannot be paid, ten days in jail is the
penalty.   If the prisoner is an "old offender" and "incorri-

gible who has appeared before the justice probably several times, he or she is sent to the "penitentiary" or the "Island" for a period not exceeding six months. To this method of dealing with intoxicated persons arrested on the street or other public place there are several objections ; in the first place, the average policeman is not a good diagnostician. Every case where the person is found stupid, dazed, or unconscious, is to him a "drunk," and must be "run in." Hence, persons suffering from stupor, partial or complete, arising from certain cerebral conditions resultant from head injury, uræmic disease, or narcotics of any kind, unless these conditions are accompanied by marked evidence of assault, or other severe injuries, are apt to be mistaken for alcoholic intoxication. This is not the fault of the police—they are not diagnosticians, neither, indeed, can be—these cases oftentimes puzzle the experienced physician. The system that allows such a state of affairs to exist is at fault, not the policeman who fails to make a proper diagnosis.

Certainly, to place such cases in a cell, and allow hours to elapse before the true condition of affairs is apprehended, is a grave and serious error.

But even if the stupor is alcoholic, and the arrest therefore legitimate, we maintain that the cell is an unfit place for such a person seriously intoxicated.

Richardson, in his "Cantor Lectures," thus writes : "Whenever we see a person disposed to meet the effects of cold by strong drink it is our duty to check that effort, and whenever we see an unfortunate person under the influence of alcohol, it is our duty to suggest warmth as the best means for his recovery.

"These facts prompt many other useful ideas of detail in our common life. If, for instance, our police were taught the simple art of taking the animal temperature of persons they have removed from the streets in a state of insensibility, the results would be most beneficial. The operation is one that hundreds of nurses now carry out daily, and applied to our police-officers at their stations, it would enable them not only to suspect the difference between a man in an apoplectic fit and a man intoxicated, but would suggest naturally the instant abolition of the practice of thrusting the really intoxicated into a *cold* and *damp* cell, which to such a one is actually an ante-room to the grave."

In view of this, in the "London Metropolitan District" the cells in which intoxicated persons are received are properly warmed in cool weather.

In addition to this we maintain that every case of alcoholic coma or stupor should come under medical supervision, that the police surgeon should make the diagnosis,— not the policeman who made the arrest—and appoint the proper remedial agencies. Too often the cell door has been shut and the prisoner allowed to "sleep off" his intoxication, and "the sleep that knows no waking" has come to him before morning.

It has been suggested that in every large city there be established a central hospital, convenient of access, where all seriously intoxicated persons, or persons found dazed or stupid upon the streets from other causes, can be taken and receive prompt medical aid, and from thence, after they are sufficiently recovered, sent to their own homes, or if friendless and homeless, assigned to the insane asylum, the inebriate asylum, the charity hospital, or such institution as seems to be most appropriate for their condition.

There is still another class, not directly coming under police supervision, to whom such central reception hospital would be a great boon—those who through alcoholic excesses develop delirium tremens or acute alcoholic delirium, those living in boarding-houses whose means are limited, who cannot command nursing and medical attention. The regular city hospitals refuse such cases, except special arrangements are made and high rates are charged ; then only cases of acute alcoholism are taken—of course, chronic alcoholics are peremptorily refused. There is good reason for this ; general hospitals have no special provision for cases of contagious disease, insane persons, or cases of alcoholism ; no padded rooms, no extra attendance, none of the appointments necessary for the care of such cases.

The suburban inebriate hospital outside of the city is already tested to its full capacity with chronic cases of inebriety. It is at some distance from the city, and to enter a patient in its wards requires certain legal formalities, and therefore, time.

Such a central hospital in the city, accessible at all times, especially to this class of cases under consideration, for which no provision is made at the general hospitals, would provide at least temporary care and treatment for insane or intoxicated persons found upon the street. It would be a channel through which the inebriate asylum, the insane asylum, or the general hospital, would receive its proper class and quota of patients. It would, as a "bureau of distribution," save much trouble now experienced in assigning insane persons to inebriate asylums, and alcoholic to insane asylums, as well as persons to either of these who might need the care of a general hospital.

While not directly established for this work, it would indirectly do considerable of it.

Besides this, the establishment of such a special hospital mainly for the treatment of such cases, would afford excellent opportunities to study alcoholism in its more acute forms. The capacity of such a hospital need not exceed fifty beds, as from it would be constantly sent out all cases not appropriate to it, and all cases assuming a chronic character.

It would not be altogether dependent on the city for its support, as the friends of many would gladly pay for the privilege of having cases treated in its wards rather than in their own homes. Acute cases of alcoholism, after recovery from the immediate attack, should their cases warrant it, could be assigned or committed to the inebriate asylum for "*chronic inebriates.*"

Fortunately, we are not without precedent in this matter. The establishment of a special city hospital devoted to the care of "acute cases of alcoholic delirium," to which the police are directed at all times to bring persons found upon the streets seriously intoxicated or stupid from other cause, and all others who may desire to have their friends treated for acute alcoholism. "The Bureau d'Admission," of the department of the Seine, at St. Anne Asylum in Paris, of which Dr. Magnan is one of the two physicians, is an institution to which no exact parallel exists in England or in this country. To it are brought all the cases of insanity previous to their admission to the various public asylums, and all cases of acute delirium or mania which fall under the care of the police in Paris. It is here that they are examined, and their admission or rejection decided upon. If admitted, they are drafted to the one or other of the asylums which is most suited to the class of the patient, or the form of his malady.

The Bureau d'Admission is quite distinct from the St. Anne Asylum itself, and under altogether different administration. In order to provide accommodation for the temporary lodgment of patients on their way to other asylums, and also for the reception of the more acute cases, it is provided with about fifty beds, and is fitted up in every way as a small asylum. Here there are brought all the cases of delirium tremens and "simple alcoholic delirium" which fall under the notice of the police, and a large number from the lower and middle classes, and here they are treated until their recovery. Hence, it comes to pass that a very large proportion of all the cases of delirium tremens occurring in Paris and its vicinity come under observation here, and this not only in one attack, but again and again, and when at last by repeated attacks they have become mentally deranged or greatly weakened, they again come under notice for transference to asylums.

The results of such opportunities of observation could scarcely fail to be productive of an increase of our knowledge, and their value is necessarily augmented by the fact of their being utilized by experienced alienists, and seen side by side with other forms of acute delirium. In addition to the hospital wards there is an out-door department, where discharged patients and others return for treatment of the various nervous disorders induced by their habits.

This hospital does excellent work, but there should be a large suburban hospital for the control of the more chronic forms of inebriety, and persons treated for an acute attack "should not be allowed to return again and again, until their minds were weakened and they became thus fit subjects for the insane asylum."

Such asylums, for long periods of commitment of chronic inebriates, exist in America, in England, in Australia, in New Zealand, and in Germany.

Much interest is now manifested by the public and the legislatures of States on this all-important topic. It is necessary, therefore, that legislative effort be directed in the proper channels, and the inebriate, who constitutes so large a proportion of our population, should be laid hold of and dealt with systematically and scientifically—not as criminals, but as those deprived of their reason and their volition—those automatic in their actions, vicious in their propensities —a curse to the community in which they dwell. Restrain, control this class, and you reduce prodigality, want, disease, to a minimum, and produce the best sort of political economy, based on science and common sense. Allow this class their liberty, and you foster these evils, and encourage and propagate their results.

Imprisonment, punishment—all punitive laws—have failed to abate or even mitigate the evil. Any effort directly based on fear of punishment or moral persua-

sion will fail. We must place the evil on its true basis,
—that of DISEASE,—and treat it accordingly. Yellow
fever, cholera, small-pox we quarantine. We investigate
the causes of epidemics, and we endeavor to remove that
cause ; so with alcohol and its attendant evils. Punitive
and restrictive laws should be directed against the manufacture and sale of alcohol, not against its victim, the
inebriate. How we shall care for the homeless, friendless,
pauper inebriate, as he is presented to us in the acute and
chronic forms of his malady, it has been the province of
this paper to point out.

CHAPTER XXVI.

SAME MEDICO-LEGAL CONSIDERATIONS.

It has become a maxim in law that drunkenness is no excuse for crime. The interpretations of phenomena by theology and medicine are undergoing modifications in consonance with the advancement in science and discovery, which distinguishes the age in which we live. Similar interpretations by the law, however, are not so impressible. The movements of the law are necessarily cautious and deliberate. The legal principle which denies to drunkenness any liberty with respect to crime, must have its reason in some presumed expediency, in the absence of exact knowledge. For the true principles which underlie the several and distinct varieties in motive and intent, inciting the inebriate in the gratification of his unnatural appetite, are even now undergoing study and analysis. The law has been unable, hitherto, to offer a comprehensive and satisfactory explanation and definition of drunkenness ; and its dictum, therefore, that inebriation is no excuse for crime, must be open to suspicion.

It is certain that a man indubitably drunk is not in his right mind, and that he can not, by any power within himself, either mental or physical, conduct himself as he would do when not intoxicated.

It is also certain that his departures from the lines of right reason are fundamental and not frivolous. But drunkenness is a state of mind and body usually of brief duration. The sober mind has means within itself of studying the nature of drunkenness between spells, as well as observing it in others. It is capable of perceiving that the use of alcoholic liquors will induce a condition of the mental faculties wherein motives and intents are unusual and unsound, as well as beyond volitional control. But the questions arise : Are there not radical differences in the motives which impel to drunkenness ? Is not intoxication, very frequently indeed, the result of the demands of a disease or of an urgency in the feelings which an attending imbecility of mind is unable to control or overcome ? If these interrogatories really fore-shadow actual facts, then it must be that there are important exceptions to the proposition that drunkenness is no excuse for crime ; for the inebriate may then be not merely irresponsible, abstractly, when drunk, but he may be irresponsible for the imbecility of will which so readily yields to the demands of the neurotic constitution.

In the neurotic constitution even slight intoxication is often succeeded by an utter blank in the memory. This withdrawal of the mind from the direct line and knowledge of conscious life implies radical disabilities in the assumption of responsibility for conduct. Drobisch explained clearly the general nature of the law of association in psychology in the language following : Psychology shows that not only memory and imagination, but judgment, reasoning, conscience itself, and, in general, all higher activity and all development of the mind rest upon the association and reproduction of states of consciousness ; that this explains also the different variations of feeling, emotion, desire, passion, and rational will. But these explications are supported by generalities that have always an indeterminate character. This arises from their lack of quantitative determination. Whatever, therefore, is conceded or permitted to a congenital infirmity of mind in its relations with the world at large, must also be accorded to constitutional incapacity in any special direction. The well-defined neurotic or spasmodic drunkard is an imbecile

in respect to his desire for intoxication ; for, in the congenital inebriate, the association and reproduction of states
of consciousness neither are, nor can they become, with
respect to his special besetment, either normal or manage-
able.

In general terms it may be said that inebriety is origin-
ally—that is, anterior to its hereditary descent in varying
forms—the outcome of very serious bodily injury, but more
especially injury to the head.   The history of the late civil
war abounds in exemplifications of this fact.   Certain physical wounds affect directly portions of the brain, or they may
withdraw from normal correspondence and relationship with
the brain to important parts of the body elsewhere.   The
means, and measure, and quality of consciousness, through
many channels of sensation and association, are thus perma-
nantly destroyed.   Thus there are produced radical defects
in consciousness, which in respect to inebriety prevent those
conservative mental operations and associations upon which
all higher activity and all development of mind rest.   I have
said neurotic inebriety is primarily occasioned by some
physical injury, possibly in remote ancestry.   This includes,
of course, such injuries to nervous integrity as may arise
from any adequate cause, perhaps not technically, yet in
reality physical, as prolonged grief, great nervous shock,
excessive study, protracted and profound disease, malaria,
and many other recognized sources of that peculiar state of
nervous instability and inadequacy which goes under the
general designation of the neurotic constitution.   Absence
of function begets incapacity to act through sheer debility
of nerve, or even through atrophy of substance.

Dr. Livingston, after years of absence amongst the black
tribes of Africa, says that upon coming into the presence of
his countrymen he was at home in everything except his
own mother tongue.   He seemed to know the language
perfectly ; but the words he wanted would not come at his
call.   It is difficult to divest the mind of the idea that the

inebriate is really capable of mastering his morbid proclivity at will. And that there may be—in view of the lack of quantitative determination in the character of the nervous disability in the neurotic inebriate—certain instances wherein the defect is not overmastering, seems probable. Yet this very qualification may doubtless include innumerable instances in which voluntary restraint is impossible. No man by taking thought can add one cubit unto his stature. Neither can a man by any process of reasoning, or any effort of will, change the functions appertaining to physical defect or pathological deterioration into the ways of normal and physiological life.

It is impossible to discuss the exact time when the brain becomes diseased by alcohol and its victim loses self-control, or what quantity of that stimulant a person can use before becoming a dipsomaniac. This point of time can be no more satisfactorily arrived at than the true time required for the production of yellow fever by the application of its exciting cause. Some constitutions would be affected in five minutes. In others it would require weeks or perhaps months of exposure to miasmata before the individual would discover the premonitory symptoms of the disease. So it is with different individuals who are in constant use of alcoholic stimulants.

It is impossible for the physician to state when the constitution is first affected by disease. The dividing line between health and disease has never been determined. Nor can it ever be defined. The physiologist has never been able to draw the dividing line between sanity and insanity, or to determine how much of the exciting cause it requires to produce a morbid condition of the brain.

These nice distinctions in regard to the pathology of disease do not enter into the discussion in reference to the importance of asylums for the control and medical treatment of dipsomaniacs.

Neither can we point out the dividing line where the

moral responsibility ceases, and the irresponsibility begins in the use of alcoholic stimulants.

The time, when an institution can reach the dipsomaniac is when he has lost self-control, and the law regards him as a dangerous citizen, or when he can be induced to enter the asylum voluntarily.

We contend that when the brain is diseased from defective nutrition, by any animal or vegetable poison, by any great shock on the nervous system impairing the nervous fluids of the body, there will be a corresponding disease of mind, which disease will develop all the peculiar types, stages, and phases of insanity, from the most inoffensive to the most furious and dangerous. It matters not how this disease may have been induced, whether by stimulants prescribed in sickness, or by the influence of social friends ; whether under extenuating circumstances, or in full view of the terrible penalty which this malady inflicts on its victim ; the State is equally bound to protect society against the insane acts of this dipsomaniac. He should be committed to an asylum for restraint and treatment adapted to his physical and mental condition.

All the laws and penalties which a State can enact against crime committed by the dipsomaniac will never prevent him, while at large, from committing murder, arson, or theft, or from taking his own life. Why, then, should our State allow its citizens to go at large, when they have lost self-control, and when daily experience shows that it is not compatible with private and public safety for them to remain at liberty ?

Does the State bring to life the murdered family by simply going through the accustomed forms of judicial procedure, in order to punish the man for what he cannot be responsible, or place him as a criminal at the bar, when his testimony would not be received in the witness box, or find out, too late, that he really is a maniac, and send him at last to an asylum as a criminal lunatic.

The only true and enlightened policy for the State is to provide asylums for this class of insane.

What is it? is the question of the hour. To find the answer, let it first be determined what or who are inebriates. That there are thousands of persons who consume intoxicants habitually and constantly, who are never visibly intoxicated, is a very obvious truth. It is nevertheless true, that there are many who, by nature and constitutional bias, are inebriates, who have never taken an inebriating draught, but who, knowing themselves, and their morbid tendency, avoid the danger of excess by absolute and perpetual abstinence.

Not a few such persons may be found among intelligent and careful people, with whom each day is a day of conflict —of conflict with themselves and their environment. They are a multitude of heroes, whose battles with self will never be known, and the record of whose conquests will never be made. Such cases have their analogues in various forms of morbid inheritance, only two of which need now be mentioned, insanity and pulmonary consumption.

The natural history of insanity and inebriety is so similiar that it is sometimes difficult to draw the line of separation. Indeed, they are so near to each other as to admit in some cases of an equal place in nosology, as, for example, in the use of the terms " insane drunkenness " and " drunken insanity," both representing kindred pathological conditions. The relationship is so patent even to the unprofessional observer, that I need dwell no longer on this point than to allude to a striking inconsistency in the law, as viewed from a medical standpoint. If I understand its meaning, the law discriminates between common drunkenness and dipsomania, but fails to recognize the likeness between dipsomania and insanity, or, in other words, it does not see a similarity between insanity from drink and insanity from other causes, though the manifestations may be similar. It assumes that the dipsomaniac is a voluntary *demon* or drunkard, and if he will, he may avoid the paroxysms that characterize the disease.

Science, however, declares a dipsomaniac, or an inebriate in the medical sense, to be what he is from an impaired or defective will, that is unable to resist the "nerve storm" which assails him at intervals, that he cannot always anticipate, as is the case in hysteria, epilepsy, etc. In consequence of this error, the law provides in the same statute for habitual drunkenness and insanity, making a criminal act committed by an insane person so far different in its results from the same act committed by an inebriate, as to warrant the commitment of the former to an asylnm, while the penalty inflicted upon the latter may be imprisonment for life in a penitentary, or it may be hanging by the neck till he is dead.

The analogy presented by pulmonary consumption may not, in the view of some, be so decided, and hence I invite attention to the early symptomatology and hereditary signs of the two disorders. Thousands are being born with a decided and well-marked consumptive diathesis, but who, knowing themselves and their family history, adapt themselves to such hygienic and climatic methods of living as tend to counteract the progress of the disease, and thus avoid its fatal ravages. Such persons, however, are consumptives by natural descent, who would go steadily on to a consumptive's lingering death, but for the knowledge of their tendency to it, and their ability to avail themselves of means to resist its approaches. While it is interesting to observe these analogues, and while the very fact of likeness serves as confirmatory evidence of disease, there is ample testimony from distinguished sources to fix the fact of disease independent of any likeness to other morbid conditions. Dr. Quain defines disease to be "any deviation from the standard of health, in any of the functions or component materials of the body."

Dr. Norman Kerr says in his recent work on inebriety,

"In drunkenness of all degrees, and every variety, the Church sees only *sin ;* the world, only *vice ;* the State, only *crime.* On the other hand, whatever else any intelligent medical practitioner beholds in such cases, he generally discovers a condition of disease." In our own country, the current medical opinion favors the same view, and I am convinced that it is gaining a firmer hold on the public mind in all departments of our social and civil life. It remains for the legal and the judicial sentiment of the land so to classify alcoholic intoxication as to remove it from the domain of morals, not even regarding it as a species of moral mania, but to accord to it its legitimate place as a physical disease.

For our present purpose, at least, it is assumed that we are agreed as to the abstract question of disease as applied to inebriety, but it becomes us to extend our inquiries a little further, that we may ascertain to what class of disorders it belongs. By common consent it is assigned to the realm of neurotic disorders. It affects most immediately and seriously the nervous system. Here, again, we may pause a moment to notice a fact concerning the complicated nervous system, which I think is not fully appreciated outside the medical profession. I refer to the sympathetic system of nerves as distinct and largely independent of the motor system. In order to apprehend the ravages of alcohol upon the sensitive nature of man, it is essential that we understand the functions of the vital, as distinct from the mechanical or automatic forces and movements of the body. I take it to be an admitted principle of law that to constitute a criminal act, the will must consent to the performance of the act, and in the study in which we are now engaged it is highly important that we discriminate as I have suggested ; that we appreciate the difference between the nervous system, which has to do with vital forces and functions, and the other nervous system, which does not control or influence vital forces or functions.

The inhibitory, restraining power resides in the series of nerves which is specifically assailed by alcohol, when taken into the body, and this is the prime fact in the whole matter of responsibility, to which the law of the land does not seem to attach importance.   I read in my Blackstone that "all the several pleas and excuses which protect the committer of a forbidden act from the punishment which is otherwise annexed thereto, may be reduced to this single consideration —the want or defect of will.   Indeed, to make a complete crime cognizable by human law, there must be both a will and an act."   In the time when this wholesome doctrine was proclaimed, the dogma of disease, as applied to inebriety, was not considered.   Intoxication was taken to be a voluntary act, and hence it was said of an inebriate that "what hurt or ill soever he doeth, his drunkenness doth aggravate it."

This doctrine may be to-day orthodox in law, but in medicine it is not, and herein lies the difference between law and medicine.   Occasions or opportunities like the present are meant to reconcile the two professions to the acceptance of this wiser doctrine and a more humane practice.

But the question arises here, and is submitted from the legal side.   If inebriety is not to be punished, how will society be protected from the assaults of the drunkard ?   Judge Davis says :   "No disease excuses any man for the commission of crime.   A man in the last stages of consumption is to be hanged for a murder as surely as though he was in perfect health, and no disease by reason of its own existence can, under any circumstances, excuse any man for the commission of crime.   Hence, to establish that it is a disease is only to put it on the exact footing on which all other diseases stand in respect of violation of law and their punishment."   If, then, insanity being considered a disease, and inebriety be taken by law to be also a disease as well defined and understood, we should gain all that we ask for.   Then I should hail this utterence from such a distinguished

source as the keynote of a new doctrine, which should be
taken up by the courts of law and sounded with accumulating force and rhythm till the jurisprudence of the whole
range of disease and crime in their joint relation shall be
infused by its healthy tone. .

While it proposes that society shall be protected from
the voluntary and deliberate criminal by punishment, be it
hanging or what else, it will protect society from the involuntary and unconscious criminal by isolation in a hospital
or asylum provided by the State for its unfortunate citizens
who have come into this world with an organization that is
out of harmony with the ethical and civil relations which
the law sanctions and provides for.   When the philosophy
of law and the science of medicine shall join hands together
to create a jurisprudence founded on such a basis, it will be
a step toward a state of society that is much to be desired,
and will be doing more in the direction of relief from the
blight of intoxication than can, in the very nature of things,
be done by the methods so ineffectually put forth at this
time.

When alcohol enters into the human body in excess, its
affinity for nerve structure is manifest in its grasp upon the
inhibitory forces as among its very early influences.   The
will is the citadel of the soul, by which life and conduct are
guarded and guided, but when it is seized and made captive,
to obey only the behests of this destructive force, the victim
is lost to himself, and acknowledges that he is enslaved.
The relation of the human will to the nervous system is
sadly misjudged.   The wonderful network of nerves known
as the sympathetic system acts independently of the will.
It presides over vital functions.   It has to do with the forces
of life with which the will has nothing to do.   The heart
beats, the stomach digests, all the vital organs fulfill their
respective offices without any reference to the will.   It
works while we sleep, and the vital functions are performed

in our unconscious rest. By its side the will is powerless. If it presides over appetite, its behests are absolute.

The inebriate, with inherited or acquired passion for stimulants, or for their hypnotic effect, cannot control his longing when it asserts itself. The hungry man who is starving for bread cannot, at his will, bid his hunger depart. The true inebriate, when his restless nerves and sinking spirits and burning desire demand repose and satisfaction, must obey the call. He obeys though his will, his conscience, his judgment, his past experiences, his moral sense, all join in earnest protest.

Is it depravity of his nature or infirmity of his will ? A wide distinction exists between depravity and pravity of will.

Depravity signifies a state of natural debasement, without any cause. The idea of a cause is precluded. It is natural.

Pravity signifies a departure from a right purpose, for which a cause is implied which is generally subjective. Indeed, an impaired, feeble will is frequently the first symptom of an approaching debauch.

CHAPTER XXVII.

MEDICO-LEGAL QUESTIONS (CONTINUED)—LAW AND RUL-INGS OF JUDGES.

In a discussion in which the question is to be considered by such able medical men from the medical side or standpoint, it has seemed to me that it would be of interest to both professions, as well as to laymen, to have the inquiry made as to those relations which attach by law to inebriety, as well in the civil and domestic relations of the inebriate, as in regard to crimes committed by persons while acting under the influence of intoxicants, or while in a state of intoxication.

What, then, is the present legal status of the question?

I shall briefly state (but have neither opportunity nor space to discuss) what I believe to be the law upon the subject; citing and grouping authorities—the civil side first, and the question of criminal responsibility second.

I. CIVIL RELATIONS. 1. *Intoxication* was regarded by the common law, when complete and characterized by unconsciousness, as a species of insanity. Lord Coke's fourth manner of "*non compos mentis*" was "4. By his own act as a drunkard."

Delirium tremens, which results directly from habits of intoxication, is in law considered to be a form of insanity, and this has been repeatedly held by the courts.

It has always been a well-settled rule of law that no person can make a contract binding upon himself while he is wholly deprived of his reason by intoxication. This would

277

be true as to deeds, wills, all instruments and obligations of every kind.

This rule is not changed where the intoxication was not procured by the other party to the contract, but is voluntary on the part of the drunkard.

By the common law, as well as by the New York statute, a testator must, at the time of the execution of a will, be of "*sound mind and memory*," and it is as requisite to have the presence of a "*disposing memory*," as a "*sound mind.*"

(*b*) By common law and by statute law an intoxicated person is thereby rendered incompetent as a witness. The statute law usually classifies such intoxicated persons as lunatics, and the provisions frequently apply similarly to each, and to both.

(*c*) In the marriage contract, which in some is treated on different grounds from all other contracts, from the necessity of the case and consequences upon consummation, the sound general rule has been, that if the party was so far intoxicated as not to understand the nature and consequences of the act, this would invalidate the contract.

2. The analogy between lunacy and total intoxication, or even habitual drunkenness, is doubtless most marked in the statutes of the various States, regarding the care and custody of the person and estates of lunatics, idiots, and habitual drunkards.

(*a*) By English law the Lord Chancellor, as the direct representative of the Crown, has always exercised the right of assuming the custody and control of the persons and estates or all those who, by reason of imbecility or want of understanding, are incapable of taking care of themselves.

Writs *de lunatico inquirendo* were issued in cases to inquire whether the party was incapable of conducting his affairs on account of habitual drunkenness.

The Supreme Court of every American State would doubtless have the right which the Court of Chancery exercised under the law of England in the absence of any statute law. This must be so in the nature of things in American States ; the principle has been exercised and adjudicated on in Kentucky, in Maryland, Illinois, Indiana, and North Carolina.

The Legislatures of the various States have vested this power by statutory enactments in various tribunals, for example in New York, by the old law in the chancellor ; in New Jersey, in the Orphans' Court ; in South Carolina, equally in the law and equity side of the courts, and now in New York, where the distinction between law and equity has been abolished, in the Supreme Court, which exercises it.

It will be observed that in many of the American States the habitual drunkard, even, is classified and treated under the same provisions, and in the same manner as the lunatic and the idiot, notably in Pennsylvania, New Jersey, Maryland, Illinois, New York, and many other States.

Taking New York as a fair illustration of the principle, it has been held by the courts, that all contracts made by habitual drunkards who have been so adjudged in proceedings *de lunatico inquirendo* are actually void. And that the disability of the habitual drunkard continues after the committee has been appointed, even when he is perfectly sober and fully aware of the nature and consequences of his acts.

It has also been held that *habitual drunkenness* being established, it is *prima facie* evidence of the subject's incapacity to manage his affairs.

We may then assume, in considering the medical jurisprudence of inebriety, that the law has always regarded and treated intoxication as a species of mental derangement, and has considered and treated the habitual or other drunkard as entitled to the special care and protection of Courts of Equity, in all matters relating to his civil rights, his domestic concerns, his ability to make contracts, his intermarrying and disposing of his property, by deed, gift, or devise.

The law has gone farther, for it has thrown round him its protecting arm and shield, when it is satisfied that he has become so addicted to drink as to seriously interfere with the care of his estate, and the courts have then come in and taken absolute control of both person and estate of drunkards, in their own interest and for their presumed good.

Medical men should keep in mind the distinction running all through the law between insanity and irresponsibility. The medical view, that irresponsibility should follow where insanity exists, has nowhere been conceded by the law, and this distinction must be borne in mind in the subject here under consideration.

II. CRIMINAL RELATIONS. This brings us to the second question : The relation of the inebriate to the criminal law for illegal acts, committed while intoxicated, which seems more harsh in its practical effect than the principles which govern him in his civil and social relations to society and the State.

This seeming hardship, however, is due to the capacity of the drunkard, considered objectively, for wrong-doing. In the one case his position as a civil agent is that of a unit of society merely—one who is, as it were, to be "saved from himself"; in the other case, the criminal aspect of the drunkard, it is the weal of society which is to be conserved and protected.

1. That form of intoxication which results in the total or partial suspension of, or interference with, the normal exercise of brain function, is regarded, at law, as mental unsoundness, and sometimes amounts to a species of insanity. It has been held at law to be a voluntary madness, caused by the willful act of the drunkard, and the decisions have been uniform, that where reason has been thus suspended, by the voluntary intoxication of a person otherwise

sane, this condition does not relieve him from the consequences of his criminal acts, or, more carefully stating it,
from acts committed by him in violation of law, while in
that state.

(a) There are decisions which go to the length of holding that the law will not consider the degree of intoxication, whether partial, excessive, or complete, and even that
if the party was unconscious at the time the act was committed, such condition would not excuse his act ; and, in
some cases, judges have gone so far as to instruct juries
that intoxication is actually an aggravation of the unlawful
act rather than an excuse.

But the better rule of law now undoubtedly is, that if
the person, at the moment of the commission of the act, was
unconscious, and incapable of reflection or memory, from
intoxication, he could not be convicted.

There must be motive and intention to constitute crime,
and in such a case the accused would be incapable from
intoxication of acting from motive.

(b) The reasons upon which the rule of law rests, may,
with great propriety, be considered, and should be carefully
studied, before any attempt at criticism is made.

1. The law assumes that he who, while sane, puts himself voluntarily into a condition, in which he knows he
cannot control his actions, must take the consequences of
his acts, and that his intentions may be inferred.

2. That he who thus voluntarily places himself in such
a position, and is sufficiently sane to conceive the perpetration of the crime, must be assumed to have contemplated
its perpetration.

3. That as malice in most cases must be shown or
established to complete the evidence of crime, it may be
inferred, from the nature of the act, how done, the provocation or its absence, and all the circumstances of the case.

In cases where the law recognizes different degrees of a
given crime, and provides that willful and deliberate inten-

tion, malice, and premeditation must be actually proved to convict in the first degree, it is a proper subject of inquiry whether the accused was in a condition of mind to be capable of premeditation.

Sometimes it becomes necessary to inquire whether the act was done in heat of passion, or after mature premeditation and deliberation, in which the actual condition of the accused and all the circumstances attending his intoxication, would be important as bearing upon the question of previous intent and malice.

(*c*) The New York Penal Code lays down with precision the provision of law governing the question of responsibility in that State as follows :

"§ 22. *Intoxicated persons.*—No act committed by a person, while in a state of intoxication, shall be deemed less criminal by reason of his having been in such condition. But whenever the actual existence, of any particular purpose, motive, or intent is a necessary element to constitute a particular species or degree of crime, the jury may take into consideration the fact that the accused was intoxicated at the time, in determining the purpose, motive, or intent, with which he committed the act."

(*d*) Voluntary intoxication, though amounting to a frenzy, has been held not to be a defense when a homicide was committed without provocation.

(*e*) *Delirium tremens,* however,—a condition which is the result of drink and is remotely due to the voluntary act of the drunkard,—has been held to be a defense to acts committed while in the frenzy, similar to the defense of insanity.

(*f*) It has been held that, when inebriety develops into a fixed and well-defined mental disease, this relieves from responsibility in criminal cases, and such cases will be regarded and treated as cases of insanity.

(*g*) It may now be regarded as a settled rule that evidence of intoxication is always admissible to explain the conduct and intent of the accused in cases of homicide.

(*h*) In crimes less than homicide, and especially where the intent is not a necessary element to constitute a degree or phase of the crime, this rule does not apply.

The practical result, however, in such cases, and in those States where the latter provision of the New York Penal Code has not been adopted, is to leave this whole subject to the judges who fix the details of punishment. This is a great public wrong, because each judge acts on his own idea, and one is merciful and the other harsh. If it is placed by law in the breast of the judges, it should be well-defined and regulated by statute. Lord MacKenzie well says : " The *discretion* of a judge is the law of tyrants."

3d. It will be observed that the law has not yet judicially recognized inebriety as a disease, except in the cases of delirium tremens—above cited—and hardly even in that case.

It is for publicists, judges, and lawmakers to consider the claim now made, that science has demonstrated inebriety to be a disease.

If this is conceded, what changes are needed to modify the law, as it at present stands, so as to fully preserve the rights of society, in its relation to the unlawful acts of inebriates, with a proper and just sense of the rights of the inebriate himself ?

The theoretical superstition that more severe punishment of inebriates will deter them from drink and crime has revived again in many sections.

A little practical investigation will show that every inebriate has a delusion that he is not a literal drunkard, but is an exception to others, and he can always stop at will when he chooses. He never realizes that any application of the law to more severe punishment will have any reference to him. He never believes that he will drink to excess or violate any law—he is not foolish enough for that. He

always deludes himself with the idea that there is no disease
in his case, and all his use of spirits is the result of accidents which he could at all times control.   Hence all
example and fear of the law are powerless.   As a lawmaker
and judge of other inebriates he is unjustly severe, but in
his own case he is always an exception, and will never come
under the general rule.   Confinement or even capital punishment of inebriates has never a personal application or is
an example in the minds of inebriates who do not suffer.
Inebriates who are sent to jail regularly every year for
intoxication always delude themselves that it is unjust and
the result of accident or personal revenge, and not of
violated law.

The inebriate who is punished for crime always consoles
himself with the faith that he is a victim of plots and conditions that should have been otherwise.

All appreciation of themselves when intoxicated is confused and cloudy, and hence he never can realize that he
will do as others have done in this condition.   The theory
of deterring these men by increased punishment has no
support practically or scientifically.   No single incident has
been produced to show that such an effect ever follows the
practical working of any law which assumes the inebriate
has the power to stop drinking, and can be forced to exercise it by intimidation and fear.

CHAPTER XXVIII.

GENERAL QUESTIONS OF IRRESPONSIBILITY.

The importance of the difficult and delicate subject of the criminal responsibility of inebriates has been considerably enhanced of recent years. The public conscience has been shocked by the severe punishment which has been inflicted on persons, for offenses committed without any criminal intention, of which offenses the doer had no remembrance when he awoke from his drunken paroxysm, and of the commission of which he was in some instances quite unconscious, in other instances impelled by a dominating narcomaniacal impulse, against which nothing short of physical restraint could have prevailed.

The marked advance made by this special question has been evidenced by the interest which has been evoked by the publication of the papers read to the Medico-Legal Society of New York, in an attractive volume, by the accomplished president, Mr. Clark Bell. The value of this book is the greater from Mr. Bell's historical account of the various papers presented to that influential and useful society (on legal relations of the inebriate in business, in social affairs, and in his responsibility before the law), during the last two decades. This interest has been increased by a thoughtful and suggestive *brochure* on the impropriety of inflicting capital punishment on inebriate criminals, by Dr. T. D. Crothers. There has been no general rule of law, all along the ages, as to responsibility for offenses complicated with intoxication. Roman law made some allowance for drunkenness, but no such consideration was exhibited in Grecian jurisprudence. Indeed, in Mitylene, under Pittacus, there was a double punishment for crimes committed while the accused was intoxicated.

In the United States, though the law recognizes no plea of responsibility on the ground of drinking, there is often manifested a practical recognition of a chronic as a diseased drunkard ; and in capital cases, the higher penalty of the law is sometimes avoided by a verdict of murder of the second degree. New York State, some fifty years ago, classed confirmed drunkards with "lunatics, idiots, and persons of unsound mind," so far as related to care of person and property. A somewhat similar classification is made in Manitoba.

In Germany and Switzerland there is a difference in the penalties for crimes committed in culpable and inculpable intoxication. By Austrian law, the accused is punished for the drunkenness only, provided he has not become intoxicated for the purpose of committing the offense. In France, the indicted inebriate is shorn of civil rights, though there is no qualification of punishment on the plea of drunkenness. In Sweden, a husband can be divorced for inebriety, and there is no mitigation of penalty from alcoholic complications. The new penal code of Italy enacts that in remission of punishment on account of intoxication at the time when the alleged criminal act is done, one-third of the sentence in money or in time is taken off.

English jurisprudence on this point is an excellent exemplification of " the glorious uncertainty of law." In the sixteenth century it was held that capital punishment must be exacted though the accused was drunk and ignorant of the fatal violence. This was re-affirmed by Lord Mansfield in the eighteenth century, he holding that drunkenness was a crime, and that one crime could not be excused by another. Coke ruled that drunkenness is an aggravation, and being an artificially contracted madness, the intoxicated madman is a *voluntarius dæmon* and there-fore responsible.

In later times judges have again and again held that drunkenness is no excuse for crime, and that a criminal act committed in a fit of intoxication is as rightly punishable as a similar act done when the accused is quite sober. For example, a man while he was drunk killed his friend who was also drunk, imagining that the latter was assaulting him violently. This prisoner was found guilty of man-slaughter on the ground that he had voluntarily become intoxicated (Reg *v.* Patterson, Norfolk Lent Assizes, 1840).

This exaction of full responsibility from a drunken ac-cused on the ground that drunkenness is a voluntary mad-ness, does not operate fairly or justly in many cases.

If drunkenness were always a voluntary act there might be something in such a contention ; but this state, in which confessedly there is often a temporary loss of reason and consciousness, is not invariably avoidable.

There are individuals who are borne involuntarily on a whirlwind of intoxication, just as at times other persons are swept off their equilibrium by a maniacal access. In the latter case, as in epileptic mania, if it can be established that the seizure is unavoidable, and the consequent actions un-controllable, complete responsibility is not exacted. In some criminal cases, complicated with drinking, the intemperate outburst during which the crime has been committed has simply been as utterly beyond the control of the person as an epileptic maniacal attack. The drunkenness has simply been a symptom of mental unsoundness. In these cases there should be no room for difference of opinion.

In other cases, though there has been no insane diathe-sis or previous insane or inebriate paroxysm, there has been a temporarily disordered nervous and mental condition which has produced a temporarily uncontrollable impulse or crave for narcotic indulgence. These morbid phenom-

ena may be the issue of a variety of unavoidable predisposing or exciting causes. For example, there is nerve
exhaustion and brain disturbance, produced in some persons
by excessive and continuous watching of a very exacting
invalid. A longer or shorter period of constant nursing
without sleep may so affect the cerebro-spinal centers that
the nurse may, for the time, be hurled into a drunken fit on
the mere sipping of an intoxicant, of which, under ordinary
healthful conditions, she could partake in limited quantities.

In the disease of narcomania (a mania for any kind of
narcotism), inherited, as in narcomania of the neurotic diathesis, there is apt to be a like risk of extreme susceptibility to the narcotic action of alcohol and other anæsthetic
intoxicants. Is it equitable that no allowance should be
made for crime committed under such circumstances ?

But over and above this inability to partake of an alcoholic or other intoxicant in limited quantity, if the smallest
sip has been tasted, there remains a still more important
phenomenon. There are many persons, who, from various
causes, operating physically and sometimes even in spite of
efforts at resistance, are impelled by an inward irresistible
impulse to rush headlong into a drunken bout. In this
transient stage of inebriate exascerbation violence may be
attempted. All admit that while drunk and beside themselves these accused are unconscious of evil intention, for the
simple reason that consciousness is for the moment practically obliterated. If such affected persons can be locked up
apart from intoxicants for a given number of hours or days
they are safe for a spell ; but unless restrained by superior
force they cannot resist the drink-impulse. Is it just that
such involuntary criminals should be punished as are voluntary evil-doers ?

Yet, again : To constitute many crimes there must be
an illegal intention. How can this be present when a man
or woman is so drunk as to be incapacitated to reason or to

remember, or even to be conscious of what he or she does ? In a recent case, where there was a sentence of twenty years imprisonment,—equivalent in the circumstances to imprisonment for life,—two men had been drinking together for hours at various bars. While at dinner in the evening, and still intoxicated, one of the drunkards shot the other. Though the judge, in his summing up, could assign no motive for the deed, the survivor of this fatal alcoholic duet was found guilty and sentenced to this heavy punishment.

Murder is sometimes done by persons who are laboring under some delusion or hallucination begotten of the narcotic brain-poisoning under which they are laboring. In one case an educated man was hung for a deliberately executed murder. Though the fact was not brought out at the trial, this victim of the law had been suffering from delusions similar to those which I have seen other persons laboring under while under the influence of chloral. But in these latter instances the patients were prevented from doing any violence by the watchful care of friends. In this class of cases human beings may suffer the highest penalty of the law for capital offenses of which they had no personal knowledge at the time, and of which they had no remembrance on emerging from the narcotic influence.

Reviewing these and many other considerations based on physical departures from health which operate to impair and for the moment destroy the moral control, and which (temporarily, it may be) so dull the consciousness that the doer of a violent deed may be either unaware of the act itself, or, if he is aware of it, his reason may be so confused that he is unable to understand the consequences or the nature of the act. Or, again, if conscious and able to understand the character and effects of the act, his will may be so paralyzed as to be powerless to resist the morbid impulse. Passing all this under review, it is most gratifying to scientific students of medical jurisprudence, to find a gradually increasing disposition in judge and jury to allow

scientific discoveries to influence their judgments.   Mr.
Justice Day, for example, recently ruled that "whatever
the cause of the unconsciousness, a person not knowing the
nature and quality of his acts, is irresponsible for them"
(Reg. v. Barnes, Lancaster Assizes, January, 1886).   If this
ruling were acquiesced in by other judges, and if juries
acted on this ruling, then a considerable proportion of
cases in which criminal offences have been committed while
the doer was in a state of drunken unconsciousness and
was therefore innocent of a criminal design, or of any actual
present knowledge of the deed, would at once be removed
from the category in which they have hitherto been almost
always placed, that of complete responsibility, involving full
penalties, and treated as irresponsible.   Indeed, the general
following of such an enlightened ruling would amount to a
revolution in our present criminal procedure.

As remarkable a judicial deliverance was that of Chief
Baron Tolles (Reg. v. M. R., Galway Summer Assizes, 1887).
The defendant, a female nurse, was accused of killing a
male patient who was under her care for typhus fever.   The
evidence showed that for over seven days she had nursed
the invalid day and night, that one night half a glass of
whiskey was given to her, and the bottle with five glasses
remaining in it was left on the kitchen dresser.   The dying
man was in charge of his mother and the nurse during the
night, when all else in the house had retired to rest.   The
mother, who was quite worn out, slept in another room,
and was awoke early in the morning by the nurse screaming.
She found her son's dead body on the kitchen floor, surrounded by fire.   The nurse was screaming and dancing
about, with a brush in one hand and a pair of tongs in the
other.   The nurse was very excited and appeared either
mad or drunk.   From other witnesses it was elicited that
the nurse cried out, "she'd soon have the devil burnt and
M. D. back again."

The judge charged that drunkenness being a voluntary

act, the law held persons responsible for acts done in a state voluntarily produced, though they did not know the nature and quality of their acts. But that, if a person, from any cause, say long watching, want of sleep, or deprivation of blood, was reduced to such a condition that a smaller quantity of stimulant would make him drunk than would produce such a state if he were in health, then neither law nor common sense would hold him responsible for his acts inasmuch as they were not voluntary but *produced by disease.* It appeared from the evidence that the nurse was under the delusion that her patient had been turned into a devil, that the proper course was to burn the devil and thus bring back the patient. Was that delusion the result of drunkenness or of disease of the mind? The jury found the prisoner guilty of manslaughter, but insane at the time of committing it, and she was ordered to be confined in a lunatic asylum during the Lord Lieutenant's pleasure.

Here again is a decision affecting a wide circle of criminal accusations. In former times the accused has suffered severe penalties in such cases, but Baron Tolles' recognition of a diseased condition and of, so to speak, an accidental involuntary intoxication as entitling to criminal irresponsibility is a remarkable event in our criminal annals.

Let me just call your attention to one more evidence of the growing influence of the discoveries of modern pathological scientific research on the judicial mind. This case is an excellent illustration of the extraordinary advance in medical jurisprudence, recognizing as it does the influence of *heredity* in modifying criminal responsibility.

An unmarried man, aged thirty-four, was charged with killing his mother, with prolonged violence, in the presence of a terror-stricken servant, whom he had locked up in the room with them all night. About five years previously the prisoner had an attack of delirium tremens, and for a year past had been subject to excited fits and delusional fears as to his life having been threatened. He persisted in declaring that the victim was not his mother. One medical witness testified that the accused was laboring under a seizure of delirium tremens when the murder was done. Another testified that he believed the form of the prisoner's illness was mania-a-potu. Evidence was adduced in proof of an insane heredity.

Baron Tollock, in his charge, said that though no man could be excused on the mere plea that he had reduced himself to want of reason by drinking, there were other circumstances in the present case. One was that through hereditary influence the accused's infirmity and mental deterioration possibly did largely account for the violent act. Another circumstance was whether, apart from drinking, the man was the subject of delusional insanity. The judge most judiciously answered the objection that if the prisoner had been an abstainer from alcoholic drink he would not have been guilty of killing his mother; that, as a certain amount of alcohol with his predisposition made him a murderer, the accused should not have taken the little drop that upset his reason. Baron Tollock replied that the last man to know his own weakness is he who has a weak mind, that such an one cannot argue as doctors can argue for him, but believes that as regards strength of mind he is on a par with all around him. The learned judge charged that if at the time when the murder was committed (though the accused had been a drunkard and had suffered from delirium tremens) he had drank only such a quantity of intoxicant liquor as an ordinary man could take without upsetting his reason, and that the insane predisposition was the main factor, although the drinking of a small quantity of alcohol was a contributary cause, the plea of irresponsibility on the ground of insanity was good. Happily the jury returned a verdict of acquittal in accordance with the judge's charge.

To these encouraging deliverances ought to be added others of an earlier date, though those were referable to a special alcoholic disease as freeing from responsibility. Though in some cases a plea based on delirium tremens has not been allowed, in other cases, such as the following, this plea has been pronounced valid. In Reg. *v.* Burns (Liverpool Summer Assizes, 1865), the accused had killed his wife, and immediately thereafter appeared to be quite calm, coolly stating that he knew what he had done, and giving as his reason for the deed that she was in league with men concealed in the walls. The jury acquitted the prisoner on the ground laid down by Baron Bramwell that, though the accused might have known that the act was killing and was wrong, he was laboring under a delusion which led him to suppose that the delusion, if true, would have justified the action.

Another person was acquitted of feloniously wounding two individuals, on the plea that he was under the impression, from delirium tremens, that his house was being broken into.

At the Liverpool Assizes, May, 1888, the jury found a verdict of " not guilty " in the case of a lady of independent fortune, on the ground that she had recently suffered from delirium tremens, had not quite recovered therefrom, and was incapable of knowing what she was doing. The alleged offence was theft of a purse, a knife, a diamond ring, and three shillings in cash.

Turn we now to our police-courts. Our present practice of dealing with drunkards there is as mischievous as it is unjust. I am informed by Dr. J. Francis Sutherland, that in my native city, Glasgow, there are some 10,000 annual commitments of intemperate women for drunkenness and offences connected therewith, for an average period of seven days. On an average each female is imprisoned three times in a year. Some forty per cent. of these prisoners have had from eleven to 800 previous convictions. What does this really mean ? Simply that, so far from curing or reforming, these short sentences actually only suffice to allow the incarcerated to recover from the effects of a " drinking bout," and send them forth once more with renewed vigor to resume their drunken excesses.

Our police-court procedure in such cases is a mere mockery of justice, a huge system for the governmental training of inebriates. All this is most unfair. When a fine is exacted the real sufferers are the children of the prisoner, who often deny themselves necessary food to gather together the amount of their parent's fine. When there is a term of imprisonment, the family are again the punished, for their means of subsistence is taken from them by the internment of the bread-winner. The latter is practically not punished, prison being but a " club " where he is provided with wholesome food, free apartments, and healthful discipline, to say nothing of gratuitous medical attendance.

Looking at our existing general criminal treatment of inebriety, there can be little doubt that it is altogether an error, and founded on a wrong conception of what drunkenness usually is. Science is day by day showing more clearly that intemperance is generally the effect of disease, the inevitable outcrop of an unhealthy condition of body or brain or both. To exact full criminal accountability from a culprit whose temporary unconsciousness and lack of control have been in the main due to certain physical perversions, temporarily or permanently affecting his reason, is as prejudicial to the individual as it is costly and demoralizing to the community, and as futile as it is unrighteous.

The legislature, by licensing the common sale of intoxicating drinks, tempts to their destruction human beings too scantily endowed with resisting-power to withstand such tremendous temptations, and having trained them in inebriety, exacts from them full responsibility for all criminal acts thus committed and developed under the *ægis* of the law. Is this intelligent, honorable, or fair ? I trow not ; and I look forward with confidence, a confidence heightened by a recollection of the remarkable tributes of the judicial bench to science, to which I have just referred, to the not far distant future, when every diseased inebriate accused of a criminal offence shall receive that fair consideration at the hands of our legal tribunals, which a righteous administration of justice owes to even the least deserving and the meanest panel at the bar.

CHAPTER XXIX.

SOME FORMS OF IRRESPONSIBILITY—ALCOHOLIC TRANCE.

The frequent statement of prisoners in court that they did not remember anything about the crime they are accused of, appears from scientific study to be a psychological fact. How far this is true in all cases has not been determined, but there can be no question that crime is often committed without a conscious knowledge or memory of the act at the time.

It is well known to students of mental science, that in certain unknown brain states memory is palsied, and fails to note the events of life and surroundings. Like the somnambulist, the person may seem to realize his surroundings and be conscious of his acts, and later be unable to recall anything which has happened. These blanks of memory occur in many disordered states of the brain and body, but are usually of such short duration as not to attract attention. Sometimes events that occur in this state may be recalled afterwards, but usually they are total blanks. The most marked blanks of memory have been noted in cases of epilepsy and inebriety. When they occur in the latter they are called *Alcoholic Trances*, and are always associated with excessive use of spirits.

Such cases are noted in persons who use spirits continuously, and who go about acting and talking sanely although giving some evidence of brain failure, yet seem to realize their condition and surroundings. Some time after, they wake up and deny all recollection of acts or events for a certain period in the past. This period to them begins at a certain point and ends hours or days after, the interval of which is a total blank, like that of unconscious sleep. Memory and certain brain functions are suspended at this time, while the other brain activities go on as usual.

In all probability the continued paralysis from alcohol not only lowers the nutrition and functional activities of the brain, but produces a local palsy, followed by a temporary failure of consciousness and memory, which after a time passes away.

When a criminal claims to have had no memory or recollection of the crime for which he is accused, if his statement is true, one of two conditions is probably present, either epilepsy or alcoholism. Such a trance state might exist and the person be free from epilepsy and alcoholism, but from our present knowledge of this condition it would be difficult to determine this fact. If epilepsy can be traced in the history of the case, the trance state has a pathological basis for its presence. If the prisoner is an inebriate, the same favoring conditions are present. If the prisoner has been insane, and suffered from sun or heat stroke, and the use of spirits are the symptoms of brain degeneration, the trance state may occur any time.

The fact of the actual existence of the trance state is a matter for study, to be determined from a history of the person and his conduct ; a grouping of evidence that the person can not simulate or falsify ; evidence that turns not on any one fact, but on an assemblage of facts that point to the same conclusion.

The following cases are given to illustrate some of these facts, which support the assertion of no memory of the act by the prisoner in court :

The first case is that of A., who was repeatedly arrested for horse stealing, and always claimed to be unconscious of the act. This defense was regarded with ridicule by the court and jury, and more severe sentences were imposed, until, finally, he died in prison. The evidence offered in different trials in defense was, that his father was weak-minded and died of consumption, and his mother was insane for many years, and died in an asylum. His early life was one of hardship, irregular living, and no training. At sixteen he entered the army, and suffered from exposure, disease, and sunstroke, and began to drink spirits to excess at this time. At twenty he was employed as a hack-driver, and ten years later became owner of a livery stable. He drank to excess at intervals, yet during this time attended to business, acting sanely and apparently conscious of all his acts, but often complained he could not recollect what he had done while drinking. When about thirty-four years of age he would, while drinking, drive strange horses to his stable, and claim that he had bought them. The next day he had no recollection of these events, and made efforts to find the owners of these horses and return them. It appeared that while under the influence of spirits the sight of a good horse hitched up by the roadside alone, created an intense desire to possess and drive it. If driving his own horse, he would stop and place it in a stable, then go and take the new horse, and after a short drive put it up in his own stable, then go and get his own horse.

The next day all this would be a blank, which he could never recall. On several occasions he displayed reasoning cunning, in not taking a horse when the owners or drivers were in sight. This desire to possess the horse seemed under control, but when no one was in sight all caution left him, and he displayed great boldness in driving about in the most public way. If the owner should appear and demand his property he would give it up in a confused, abstract way. No scolding or severe language made any impression on him. Often, if the horse seemed weary, he would place it in the nearest stable, with strict orders to give it special care. On one occasion he joined in a search of a stolen horse, and found it in a stable where he had

placed it many days before.   Of this he had no recollection.
In another instance he sold a horse which he had taken,
but did not take any money, making a condition that the
buyer should return the horse if he did not like it.   His
horse stealing was all of this general character.   No motive
was apparent, or effort at concealment, and on recovering
from his alcoholic excess, he made every effort to restore
the property, expressing great regrets and paying freely for
all losses.   The facts of these events fully sustained his
assertion of unconsciousness, yet his apparent sanity was
made the standard of his mental condition.   The facts of
his heredity, drinking, crime, and conduct all sustained
his assertion of unconsciousness of these events.   This was
an alcoholic trance state, with kleptomaniac impulses.

The next case, that of B., was executed for the murder
of his wife.   He asserted positively that he had no memory
or consciousness of the act, or any event before or after.
The evidence indicated that he was an inebriate of ten years
duration, dating from a sunstroke.   He drank periodically,
for a week or ten days at a time, and during this period was
intensely excitable and active.   He seemed always sane and
conscious of his acts and surroundings, although intensely
suspicious, exacting, and very irritable to all his associates.
When sober he was kind, generous and confiding, and never
angry or irritable.   He denied all memory of his acts during
this period.   While his temper, emotions, and conduct were
greatly changed during this time, his intellect seemed more
acute and sensitive to all his acts and surroundings.   His
business was conducted with usual skill, but he seemed unable to carry out any oral promises, claiming he could not
recollect them.   His business associates always put all bargains and agreements in writing when he was drinking, for
the reason he denied them when sober.   But when not
drinking his word and promise was always literally carried
out.   He broke up the furniture of his parlor when in this
state, and injured a trusted friend, and in many ways showed

violence from no cause or reason, and afterwards claimed no memory of it.

After these attacks were over, he expressed great alarm and sought in every way to repair the injury. Finally he struck his wife with a chair and killed her, and awoke the next day in jail, and manifested the most profound sorrow. While he disclaimed all knowledge of the crime, he was anxious to die and welcome his execution. This case was a periodical inebriate with maniacal and homicidal tendencies. His changed conduct, and unreasoning, motiveless acts, pointed to a condition of trance. His assertion of no memory was sustained by his conduct after and efforts to find out what he had done and repair the injury.

The third case, that of C., was a man of wealth and character, who forged a large note, drew the money and went to a distant city on a visit. He was tried and sentenced to State prison. The defense was no memory or consciousness of the act, by reason of excessive use of alcohol. This was treated with ridicule. Although he had drank to excess at the time and before the crime, he seemed rational and acted in no way as if he did not understand what he was doing. Both his parents were neurotics, and he began to drink in early life, and for years was a moderate drinker. He was a successful manufacturer, and only drank to excess at times for the past five years. He complained of no memory during these drink paroxysms, and questioned business transactions and bargains he made at this time. On one occasion he went to New York and made foolish purchases which he could not recall. On several occasions he discharged valuable workmen, and when he became sober took them back, unable to account for such acts.

These and other very strange acts continued to increase with every drink excess. At such times he was reticent and seemed to be sensible and conscious, and did these strange acts in a sudden, impulsive way. The forged note was

offered boldly, and no effort was made to conceal his presence or destination. When arrested, he was alarmed and
could not believe that he had done so foolish an act. This
was a clear case of alcoholic trance, in which all the facts
sustained his assertion of no conscious memory of the crime.
In these three cases the correctness of the prisoner's assertions of no memory was verified by all the facts and circumstances of the crime. The mere statement of a person
accused of crime, that he had no memory of the act, should
lead to a careful examination and be only accepted as a
fact when it is supported by other evidence.

The following case illustrates the difficulty of support-
ing a prisoner's statement of no memory when it is used for
purposes of deception :

Case E. An inebriate killed a man in a fight, and was
sentenced to prison for life. He claimed no memory or
recollection of the act. I found that when drinking he
seemed conscious of all his surroundings, and was always
anxious to conceal his real condition, and if anything had
happened while in this state he was very active to repair
and hush it up. He was at times quite delirious when
under the influence of spirits, but would stop at once if any
one came along that he respected. He would, after acting
wildly, seem to grow sober at once, and do everything to
restore the disorder he had created. The crime was an
accident, and at once he attempted concealment, ran away,
changed his clothing, and tried to disguise his identity ;
when arrested, claimed no memory or consciousness of the
act. This claim was clearly not true, and contradicted by
the facts.

In a recent case, F. shot his partner in business while both
were intoxicated, and displayed great cunning to conceal
the crime and person ; then, after elaborate preparations,
went away. He made the same claim of defense, which was
unsupported by any other evidence or facts in his previous
life. He was executed. Of course it is possible for the

trance state to come on suddenly, and crime be committed at this time ; still, so far, all the cases studied show that this condition existed before, and was the product of a growth beginning in brief blanks of a few moments and extending to hours and days duration. Unless the facts indicated the trance state before the crime was committed, it would be difficult to establish this condition for the first time, followed and associated with the crime.

I think in most of these cases, where this defense is set up, there will be found certain groups of cases that have common physical conditions of degeneration. These groups of cases I have divided from a clinical standpoint, the value of which will be more as an outline for future studies.

Probably the largest number of criminal inebriates who claim loss of memory as a defense for their acts, are the alcoholic dements. This class are the chronic inebriates of long duration ; persons who have naturally physical and mental defects, and who have used spirits to excess for years. This, with bad training in early life, bad surroundings, and bad nutrition, have made them of necessity unsound, and liable to have many and complex brain defects. Such persons are always more or less without consciousness or realization of their acts. They act automatically only, governed by the lowest and most transient impulses. Crimes of all kinds are generally accidents growing out of the surroundings, without premeditation or plan. They are incapable of sane reasoning or appreciation of the results of their conduct. The crime is unreasoning, and general indifference marks all their acts afterwards. The crime is always along lines of previous conduct, and never strange or unusual. The claim of no memory in such cases has always a reasonable basis of truth in the physical conditions of the person. Mania is very rarely present, but delusions and morbid impulses of a melancholic type always exist. The mind, like the body, is exhausted, depressed, and acts along lines of least resistance.

The second group of criminals who claim no memory are those where the crime is unusual, extraordinary, and unforeseen. Persons who are inebriates suddenly commit murder, steal, or do some criminal act that is foreign to all previous conduct. In such cases the trance condition may have been present for some time before and escaped any special notice, except the mere statement of the person that he could not recollect his acts. The unusual nature of the crime, committed by persons who never before by act or thought gave any indication of it, is always a factor sustaining the claim of no memory. The explosive, unreasoning character of crime always points to mental unsoundness and incapacity of control.

A third group of criminals urge this statement of no memory, who, unlike the first group, are not imbeciles, generally. They are positive inebriates, drinking to excess, but not to stupor, who suddenly commit crime with the most idiotic coolness and indifference, never manifesting the slightest appreciation of the act as wrong, or likely to be followed by punishment. Crime committed by this class is never concealed, and the criminal's after conduct and appearance gives no intimation that he is aware of what he has done. These cases have been termed moral paralytics, and the claim of the trance state may be very likely true.

A fourth group of cases, where memory is claimed to be absent, occurs in dipsomaniacs and periodical inebriates, who have distinct free intervals of sobriety. This class begin to drink to great excess at once, then drink less for a day or more, and begin as violently as ever again. In this short interval of moderate drinking some crime is committed which they claim not to have any recollection.

Other cases have been noted where a condition of mental irritation or depression preceded the drink explosion, and the crime was committed during this premonitory period and before they drank to excess. The strong probability of trance at this period is sustained by the epileptic character of such conduct afterwards. , The trance may be justly termed a species of *aura*, or brain paralysis, which precedes the explosion.

In some instances, before the drink storm comes on, the person's mind would be filled with the most intense suspicions, fears, delusions, and exhibit a degree of irritation and perturbation unusual and unaccountable. Intense excitement for depression, from no apparent cause, prevails, and during this period some crime may be committed ; then comes the drink paroxysm, and later all the past is a blank. Trance is very likely to be present at this time.

In these groups the crime is generally automatic, or committed in a manner different from other similar crimes. Some governing center has suspended, and all sorts of impulses may merge into acts any moment. The consciousness of acts and their consequences are broken up. The strong probability is that these trance blanks begin in short periods of unconsciousness, which lengthen with the degeneration and mental feebleness of the person. The obscurity of these conditions, and the incapacity of the victims to realize their import, also the absence of any special study, greatly increases the difficulty. It will be evident from inquiry that trance states among inebriates are common, but seldom attract attention, unless they come into legal notice.

The practical question to be determined in a given case in court is the actual mental condition of the prisoner, who claims to have no recollection of the crime. This is a class of evidence that must be determined by circumstantial and collateral facts, which require scientific expertness to gather and group. The court can decide from the general facts of the crime and the prisoner whether his claim of no memory may possibly be true, and order an expert examination to ascertain the facts. This should be done in all cases where the prisoner is without means, in the same way that a lunacy commission is appointed to decide upon insanity.

The result of this expert study may show a large preponderance of evidence sustaining the claim of no memory, or the opposite. If the former, the measure of the responsibility must be modified, and the degree of punishment changed. While such cases are practically insane at the time, and incapable of realizing or controlling their acts, they should be kept under legal and medical surveillance for a lifetime, if necessary. Such men are dangerous, and should be carefully watched and deprived of their liberty for a length of time depending on recovery and capacity to act rationally and normally. They are dangerous diseased men, and, like victims of contagious disease, must be housed and treated.

The future of such cases depends on the removal of the causes which made them what they are. The possibility of permanent restoration is very promising in most cases. How far alcoholic trance exists in criminal cases is unknown, but the time has come when such a claim by criminals cannot be ignored, and must be the subject of serious inquiry. Such a claim cannot be treated as a mere subterfuge to avoid punishment, but should receive the same attention that a claim of insanity or self-defense would. This is only an outline view of a very wide and most practical field of medico-legal research, largely unknown, which can be seen in every court-room of the land. These cases appeal to us for help and recognition, and the highest dictates of humanity and justice demand of us an accurate study and comprehension of their nature and character.

The following summary of the leading facts in this trance condition will be a standpoint for other and more minute investigations :

1st. The trance state in inebriety is a distinct brain condition, that exists beyond all question or doubt.

2d. This brain state is one in which all memory and consciousness of acts or words are suspended, the person going about automatically, giving little or no evidence of his real condition.

3d. The higher brain centers controlling consciousness are suspended, as in the somnambulistic or hypnotic state. The duration of this state may be from a few moments to several days, and the person at this time may appear conscious and act naturally, and along the line of his ordinary life.

4th. During this trance period crime against person or property may be committed without any motive or apparent plan, usually unforseen and unexpected. When accurately studied such a crime will lack in the details and methods of execution, and also show want of consciousness of the nature and results of such acts.

5th. When this condition passes away the acts and conduct of the person show that he did not remember what he had done before. Hence his denial of all recollection of past events and his changed manner confirm or deny his statements.

6th. When such cases come under judicial inquiry the statement of the prisoner requires a scientific study before it can be accepted as a probable fact. It cannot be simulated, but is susceptible of proof beyond the comprehension of the prisoner.

7th. In such a state crime and criminal impulses are the result of unknown and unforeseen influences, and the person in this condition is dangerous and an irresponsible madman.

8th. This condition should be fully recognized by court and jury, and the measure of responsibility and punishment suited to each case. They should not be punished as criminals, nor should they be liberated as sane men. They should be housed and confined in hospitals.

CHAPTER XXX.

PRACTICAL DIFFERENTIATION OF INEBRIETY FROM COMA, ETC., ETC.

The frequent occurrence of blunders in mistaking brain-diseases for drunkenness, and the serious reproach they bring on medical men, render it necessary that more earnest attention should be paid to the subject than heretofore, and that a higher knowledge should be obtained of the character of the dangers incident to these accidents. Unfortunately, drunkenness has not, save in a few instances, been studied as a disease, and consequently the manifestations pertaining to it are very little understood. The ignorance is particularly unfortunate when it is necessary to distinguish between it and brain-troubles.

In starting out in this discussion, our first duty will be to enumerate the different conditions which may be mistaken for drunkenness, and the symptoms of which it is necessary to bear in mind in forming a diagnosis.

These are:

1. Fracture of the skull.
2. Concussion of the brain.
3. Cerebral hemorrhage.
4. Embolism and thrombosis.
5. Uræmia, from Bright's disease.
6. Epilepsy.
7. Narcotic poisoning.
8. Heat apoplexy.

We will take up these lesions one by one.

In cases of fracture of the skull, or where severe or stunning blows have been dealt, the greatest difficulty is met with in the diagnosis in the absence of any history of the case, for the reason that the coma in these instances is frequently profound, and simulates that of drunkenness. The smell of the breath should never be relied on as a test, for many industrious and useful workmen are in the habit of taking a certain amount of liquor during the day.

The temperature, the condition of the pupils, the breathing, should all be carefully observed, but the true rule is to keep the patient under close and constant watch, until a fixed diagnosis is obtained. It is also important, in these cases, to look closely for wounds and marks of violence.

Mr. Lawson, of Middlesex Hospital, relates an interesting case bearing on this point. " The patient was taken to the police-cell as drunk. He was medically examined, and recovered sufficiently from his apparent drunken semi-consciousness to be able to converse with those about him. After a few hours, however, severe cerebral symptoms came on, and he was transferred to the hospital, where he died on the thirteenth day from severe lacerations of the brain-substances, associated with extensive hemorrhage, and with fracture into the lamboidal suture. A remarkable point in this case was the absence of paralytic symptoms, considering the severe laceration of the brain. With the exception of the loss of power over the sphincters, there was no paralysis whatever."

Cerebral hemorrhage is more frequently mistaken for drunkenness than any other trouble, for the reason that the symptoms are similar in several stages of the two diseases. There is a stage of noisy violence and uproar in both, and also a condition of complete coma. In ordinary cases of apoplexy we look for paralysis of one side or the other, but this does not always obtain, if the hemorrhage be into the pons, or lateral ventricle. We may have convulsions in both diseases, but usually they are more severe on one side of the body in apoplexy.

The state of the pupils cannot always be relied on as a differential test, although squinting as well as conjugate deviation of the eyes is a distinctive mark of apoplexy. Doctor MacEwen, of Glasgow, says the ordinary opinion that dilation of the pupils is found in alcoholic coma is incorrect, but that contraction is the rule. He accidentally discovered, however, that if a patient was shaken or disturbed, the pupils dilated, but very soon contracted again. He therefore lays down the rule that an insensible person who, being left undisturbed for from ten to thirty minutes, has contracted pupils, which dilate on his being shaken, without any return of consciousness, and then contract again, can be laboring under no other state than alcoholic coma. Strange to say, Dr. Reynolds has witnessed the same phenomena in patients suffering from acute softening. He says in his system of medicine, that he has often raised the lids of patients in this condition, and exposed the contracted pupils to the light without arousing them ; that there is no dilation or change to be observed, but if they be addressed loudly by name, or their toes pinched so that they awake, the pupils instantly dilate.

The truth is, that in cerebral hemorrhage the pupils present no fixed peculiarity. There may be a clot on one side of the brain, and yet the pupils appear normal. Their condition may even vary in different cases of the same lesion. Cases of ingravescent apoplexy generally commence with delirium or convulsion, and the coma comes on slowly and gradually. These are the cases that are frequently mistaken for drunkenness, provided the smell of alcohol be discovered on the breath of the patient.

Cases of embolism and thrombosis should not be confounded with drunkenness. In embolism, the coma is sudden and transient, and in thrombosis, the paralytic symptoms are so marked that an error can scarcely occur. Fatal cases of sudden coma and paralysis, with partial recovery

of consciousness and power, are met with, independent of
drunkenness or brain trouble.

A remarkable case of this character is reported in the
Lancet, in which the only lesion found after death was
hydatids of the pineal gland, liver and peritoneum. The
patient died in thirteen hours from the commencement of
the attack. In cases of coma from uræmia, the diagnosis is
not so difficult, inasmuch as we have some well-marked
points for our guidance. This form of coma is generally
preceded by convulsions. The breath has a peculiar unmistakable fetor, and the urine upon examination will be
found to contain a large quantity of albumen as well as
other deposits, indicating kidney disease. Occasionally,
however, cerebral hemorrhage is present along with uræmic
poisoning, and this complicates to some extent the diagnosis.

Another difficulty in diagnosis in that the urine may
become temporarily albumenous from the inordinate use of
alcohol. Dr. George Johnson mentions a case of this character, which occurred in the practice of his friend Dr. Baxter.
" A man between twenty and thirty years of age was brought
in one night by the police. He was unconscious and breathing stertorously. He appeared to be drunk, and a large
quantity of vinous liquid was pumped out of his stomach.
The unconsciousness continued, and it was then suspected
that he might be suffering from uræmic poisoning. This
suspicion was confirmed by the fact that his urine, drawn
off by a catheter, was loaded with albumen. He was then
put to bed, cupped over the loins, and a purgative was given.
When Dr. Baxter visited the ward next morning, he found
the man up and dressed, and clamoring for his discharge.
He said he had been very drunk over night, but now he
had nothing the matter with him ; and he passed some
urine, which was found to be in every respect quite normal.
The temporary albuminuria was the result of renal congestion, caused by the excretion of an excess of alcohol through
the kidneys."

Epilepsy can generally be diagnosed without much difficulty, though if the patient is not seen during the attack, but only during the profound and prolonged coma which sometimes follows the paroxysm, the case may be mistaken for one of drunkenness. Epileptic coma, however, is usually of short duration, and if the tongue is bitten or bleeding, or if hemorrhagic spots be discovered beneath the conjunction or skin, all uncertainty with regard to the case will be cleared up.

The coma resulting from the poisoning by opium is very similar to that produced by the administration of large quantities of alcohol. At one time it was believed that extreme contraction of the pupils was a distinguishing mark of coma resulting from opium, but this, it is known, cannot be relied on, inasmuch as the pupils are often found contracted, as I have before mentioned, in alcoholism, and also, as mentioned by Dr. Wilks, in apoplexy seated in the pons varolii. In cases of opium poison that I have seen, I have always thought that the breathing was much slower than in the coma produced by drunkenness, but in this I may be mistaken. The smell of opium is frequently to be discovered on the breath, particularly if laudanum has been taken, and this becomes an important feature, provided no history of the case can be obtained.

There are many symptoms in severe cases of heat apoplexy which might mislead an inexperienced practitioner, and cause him to believe his patient is suffering from the effects of drunkenness. Coma is very often the result of sunstroke, and great mental disturbance and outward violence are not unfrequent results of aggravated cases. I can remember several instances in which I was in doubt for a time in regard to the origin of the symptoms present in cases which afterwards proved to be heat apoplexy. There is one very simple diagnostic mark in sunstroke which is never absent, and which will greatly aid us in forming a judgment : it is intense heat of the head—a heat which is to be found in no other disease, save yellow fever. In the coma of drunkenness this extreme heat is never found—at least, I have not met it in my own experience.

I have seen many cases of hysteria in women which at first puzzled me, inasmuch as the symptoms were similar to those induced by the action of alcohol. These cases are not so much characterized by coma as by a state of excitement and violent demonstration. The phases of hysteria are so varied, and the abnormal manifestations of this trouble so curious, that the ordinary practitioner may readily mistake neurotic trouble for the effects of alcohol ; as, on the contrary, he may and does frequently mistake drunkenness for hysteria. The puzzling cases I have met with are those in which there was a combination of whiskey and hysteria, a condition which, I may add, is of the most delightful character, and affording a train of the most original and beautiful manifestations. The methods of examination in coma may be summarized as follows :

First. An examination of the head and body for fracture of the skull and external injuries.

Second. Examination for hemiplegia, squinting, conjugate deviation or facial paralysis.

Third. Examination of the mouth and tongue.

Fourth. The legs and eyelids should be examined to see if œdema had previously existed.

Fifth. An examination of the urine both for albumen and alcohol.

Sixth. A stomach pump may be used in many cases with great advantage.

Seventh. A history of the attack and its general features should be inquired into, if possible.

Eighth. The pupils and breath should both be examined and the temperature taken, though undue importance, as before stated, should not be attached to the evidences afforded by such an examination.

In conclusion, let me particularly dwell on the importance of close attention and watchfulness in all cases of coma supposed to be due to drunkenness. The system heretofore pursued has been most barbarous, both in this country and Europe, and is a reproach to our civilization. Dwelling on this subject, Dr. John Curnow pointedly says : " I must enter a protest against the routine treatment of drunkenness too generally followed, viz.: emetics or the stomach pump, cold effusion, flecking the skin with a wet towel, and then the interrupted galvanic current. A patient having grumbled out a name, and perhaps an address, is turned over to a policeman, who speedily consigns him to a cold cell to sleep off his symptoms. It cannot too often be insisted upon that a drunken man is suffering from acute poison, and cannot be too closely watched."

All police stations should have a regularly appointed medical officer in charge, and every case of sickness, or aggravated case of drunkenness, should be put under his care. Certain instruments and appliances should be constantly at hand, and supplied at public expense, such as a stomach pump, galvanic battery, hypodermic syringe, test tubes, cupping apparatus, as well as mustard, apomorphia, etc. When these precautions are taken, and when inebriety is added to the list of diseases, and its treatment taught in our schools, many lives will be saved and much unhappiness be spared to the community.

CHAPTER XXXI.

GENERAL CONSIDERATIONS OF OPIUM INEBRIETY.

Opium neurosis is not an intoxication from the drug, but a central neurotic change, brought about by the long persisting perversion of function and impairment of central nervous nutrition, from its persisting presence in the nutrient pabulum of the circulation.

The psychosis of opium is a blended intoxication and chronic poisoning of the psychical centers of the brain ; other symptoms of acute opium poisoning are essentially different, being ·mainly a profound paralysis of sensation and of the centers of involuntary motion especially having their origin in the medulla and upper part of the spinal cord—profound narcosis, lowered respiratory movements, etc., while chronic opium poisoning, or meconeuropathia, is characterized by repeated nerve excitations, in which the nerve centers, not being completely overcome, a kind of tolerance is established, with progressively developing abnormal molecular neural changes, which are as repeatedly covered up and masked by the renewed doses, till some sudden deprivation of the drug or failure to appropriate it, reveals, in full force, the neural mischief which has been gradually done. Opium, like a bank defaulter, both makes and masks the mischief done, which may be kept concealed so long as he stays in the institution.

Beyond all question, the toxic use of opium and its alkaloids is rapidly increasing. Only about 50 per cent. of opium and morphine manufactured is required by the legitimate demands of medicine and pharmacy. The enormous balance is consumed in some unknown way. Comparative estimates make the number of opium cases in this country over a hundred thousand. Whether this is correct or not, it is evident that the number is very great and largely concealed, and many of them are very hopeless and difficult to treat. The natural history of such cases indicates a steady, progressive degeneration on to death. Recovery is rarely spontaneous and without the aid of applied science.

The central tracts involved are the cerebro-spinal and sympathetic systems. Deviations from health noted are due to departure from the normal tone of one or both of these centers. Organic lesions are rare, possibly some instances of renal or brain disease, the usual ultimate result being a state of marasmus, impaired nutrition and profaned nerve depression ending in death.

From a careful clinical study and grouping of the history of a number of opium cases, it is evident that a large proportion have a distinct *neurotic diathesis*, or, more literally, have inherited from their parents some condition of brain and nerve defect which favors and predisposes to the development of neurotic diseases. A more careful study of these records shows that in some cases an *opium diathesis* is present, or a special inherited tendency to use opium. Here are two conditions which influence and favor this disease. It is a well-known fact that a large proportion of all nerve and brain diseases appears in children of neurotic and defective parents. Such children have received some special tendency and predisposition favoring the growth of nerve diseases, springing into activity from the slightest causes.

The latency or activity of this diathesis will depend on certain conditions of life and surroundings, which in many cases can be traced. In some instances the diseases of parents reappear in the children, in others in allied diseases, and not infrequently these defects pass over and reappear in the third generation. Often such defects are dormant, and only break out from the application of some peculiar exciting cause. Thus a hysteric mother and paranoic father were followed by three children. One was an alcoholic, the second was a wild, impulsive temperance reformer, the third was a sad, depressed, melancholic man. In the third generation opium and alcoholic inebriety, insanity, pauperism, also feebleness of mind and body, appeared. These varied forms of nerve diseases all had a neurotic diathesis as a basis, and the different phases were the direct result of different exciting causes. These facts are numerous and well attested, and so uniform in their operation that it is entirely within the realm of possibility to predict that, from a knowledge of the diseases of the parents and the environment of the child, certain forms of degeneration and diseases will appear with almost astronomical precision. This term, " neurotic diathesis," covers a vast unknown field of causes which extend back many generations.

The evolution of brain and ner⁄ ᵗ defects can often be traced through the realms of environment, nutrition, growth, and development. Medical text-books and teaching which fail to recognize this, give very narrow conceptions and strange exaggeration of the influence and force of many insignificant and secondary factors in the production of disease. The opium-taker has often this neurotic element in his history. It may be traced back to his ancestors, or it may be associated with brain or nerve injuries, cell-starvation, faulty nutrition, auto-intoxications, brain strains, or excessive drains of nerve force. A train of predisposing causes may have been gathering for an indefinite time back. Then comes the match which kindles or fires the train of *gathering forces*. This same train of exciting causes may not explode, because the germ soil is absent. Opium in all forms is given daily, and yet only a comparatively small number of cases become addicted to its use. Why should an increasing number of persons take opium continuously for the transient relief it gives? Why should the effects of this drug become so pleasing as to demand its increased use, irrespect-

ive of all consequences ?   The only explanation is the presence of a neurotic diathesis, either inherited or acquired.

There is a large class of opium cases in which a complex
diathesis exists—particularly following inebriety and various forms of brain exhaustion   Often alcoholics will use
opium irregularly and transmit to their descendants a
diathesis which very commonly favors the use of this drug.
Thus the alcohol diathesis frequently becomes the opium
craze, with but slight exposure.   Both of these disorders
are rapidly interchangeable.   The children of opium takers
may turn to alcohol for relief, and *vice versa*.   It is clear
that the moderate use of alcohol produces a degree of
degeneration that frequently appears in the next generation
as predisposing causes to the opium or allied diseases.
Clinical study of cases brings ample confirmation of this.
The children of both alcohol and opium inebriates display
many forms of brain degeneration.   The paranoics, criminals,
prostitutes, paupers, and the army of defects, all build up a
diathesis and favoring soil for the opium craze.   Descendants from such parents will always be markedly defective.
They are noted by brain and nerve instability, hyperæsthesia,
and tendency to exhaustion ; also extreme pain from every
degree of functional disturbance, with low powers of restoration, inability to bear pain, and suffering from mental
changeability, impulsiveness and drug credulity, etc.

These characteristics are prominent, and mark a neurosis that quickly merges into the opium disease.   Yet a min-
ority of these cases show a sensitiveness in the effects of
opium that prevents them from using it.   I have seen a
neurotic patient become dangerously narcotized by the use
of half a grain of solid opium.   Some of the alcoholics and
other narcomaniacs have exhibited an incompatibility to
opium that is often startling.   The emesis and prostration,
and the brain stimulation which approaches and becomes
hyperæmia from one or more doses, are familiar to all.
This intolerance precludes the use of the drug, and is recog-

nized with alarm by the patient. On the other hand, when
the effects are rapid and marked, relieving pain or restoring
the disturbance of the functions, with no other than a pleasing sense of rest and cure, a dangerous diathesis should be
suspected.   While the physician recognizes the constitutional incompatibility in one case, he ought not to overlook
the abnormal attractiveness of the drug in the other. The
dose of morphine which gives the first complete rest, or
calms the delirious excitement, or relieves the neuralgic
pain or the digestive disturbance, soon calls for its repetition, and many physicians will unconsciously sanction and
advise its use. Thus, far more fatal conditions are culti-
vated and roused into activity.

In all neurotic cases, the use of opium in any form when
given, should be concealed and watched with care. If a
special predilection for this drug appears, equal care and
skill should be exercised to divert and change it. Opium
should only be used from a knowledge of the nature and
character of the case. I have seen the most disastrous
results from the reckless use of morphine with the needle.
Recently, a man to whom morphine was intolerant was cut
and stunned by a falling plank in the street. The surgeon
gave him a hypodermic of morphine and ordered him to the
hospital. He died in a short time from opium neuroses.
Police surgeons often make this mistake, giving morphine
that from some unknown reason becomes fatal.

There is another class of opium takers in which abnormal nutrition seems to be the most active factor in the causation. The neurotic or opium diathesis is not apparently
present, and opium-taking dates from some nutrient disturbance. Such cases are very commonly sufferers from dyspepsia, derangement of the liver and bowels. They have a
deranged appetite, headaches, cramps, thirst, and fever at
times, with nausea. They are anæmic and hyperæsthetic
and complain of varied pains and neuralgia. These cases
are evidently ill-nourished, and, in all probability, suffer from

imperfect digestion, assimilation, and elimination of food-
products and waste material. Poisonous compounds and
auto-intoxications form sources of serious trouble. The
brain suffers from fatigue and pain, the cells are imperfectly
nourished, and congestions, complex neuralgias, nerve irritation and instability follow. For this condition opium is
almost a specific paralyzant. These cases are found among
the over-fed, the under-fed, and those who neglect common
hygienic rules of living.

Cases of the over-fed are usually epicures, gormands,
and persons living sedentary lives, and eating at all times
and places. Dyspepsia and derangement of the bowels and
kidneys make them drug-takers; then follows opium in some
form. Defective elimination and auto-intoxications are
always present. The under-fed are usually misers, or persons
very poor and very neglectful of themselves, or paranoics
who have some food delusion. They are practically suffering from cell and tissue starvation and nutrient debility.
The same dyspepsia and bowel derangements follow. Then
follows drug-taking or special foods, and soon opium is discovered and adopted as a remedy. The same poisonous
waste products appear from deranged assimilation; also,
elimination and the nerve centers are deranged by these
new and dangerous chemical compounds. The class of
persons who, from simple neglect, become diseased, are often
the very poor and ignorant, or some division of the great
army of border-liners, who live both mentally and physically
on the very frontiers of sanity and insanity. Such persons
clearly suffer from many and various forms of auto-intoxications, and this is proven inductively by the result of
eliminative treatment.

In all of these cases of nutrient neglect, many favoring
conditions encourage the use of opium. These cases are
numerous and comprise a large part of the invalids,
hypochondriacs, and chronic drug-takers, who are seen in
our offices and at the dispensaries. They are all practically

suffering from faulty assimilations, and faulty eliminations and the irritation of retained poisonous compounds. Opium is a remedy of positive force in covering up the protests of the defective cells and irritable nerves. Often these cases are concealed, and are partly the result of previous disorder, and partially acquired from the effects of opium.

Next to this class of nutrient sufferers who become opium-takers, are those who have some entailment of disease or injury. In their history it will appear that some stage of invalidism was present, dating from brain, nerve, or bodily injury. Fevers, heat, or sun-strokes, brain shocks from any source, which are followed by unconsciousness or marked mental perturbations, with exhaustion, and also a profound lowering of all the vital forces. These and other events have left damaged functional and organic activities, manifest in various neuralgias and physical disturbances.

The use of opium conceals and covers up this trouble. Many veterans of the late war have become opium maniacs for the relief of their pains and sufferings, and this is often concealed where it might possibly peril the procuring of a pension. The pension bureau should recognize the use of opium as a natural sequence and entailment following the disease and injury in the service. In Prussia, both alcohol and opium inebriety are treated as diseases when occurring in the army or civil service. The suffering and hardships growing out of the war has been the exciting cause of a great many opium cases. Many persons who have no special nerve diathesis in their history, after some severe illness, injury, or mental strain, exhibit a degree of nerve instability and feebleness that is significant of serious organic change. Such persons manifest perversions of taste, with delusions of food and medicines, and are on the border-lines of narcomania, ready to use any food or drugs which will bring even transient relief.

The use of opium is always perilous. Why all these and

similar cases do not become opium-takers is owing to the absence of some diathesis inherited or acquired.

We can see some of the many complex causes favoring brain and nerve strain, with rapid exhaustion and degeneration, and the interchangeability of nerve diseases, in which the use of opium is only another form of the same disease. But we cannot yet trace the early causes and cell-conditions which develop the opium craze. This morbid impulse, like the delirious thirst for water on a desert plain completely dominates all reason and so-called will-power, and every consideration of life and surroundings. It is more than an accident, more than a failure to reason and act wisely ; it is a disease, an organized march of dissolution. The demand for opium is only a symptom ; the removal of opium is not the cure. Some central brain degeneration has begun and is going on. Narcomania, a morbid thirst for any solids or fluids that will produce neuroses, is the general name, and opium mania is only one member of this family.

In this study the fact is emphasized that the opium disease appears most frequently in persons who have a neurotic and opium diathesis, and also in persons who are suffering from nutrient disturbances, and those who are invalids or have some entailment of previous disease and injury ; also that certain diseases and symptoms seem to furnish favoring soils for its growth and development. While these are but faint outlines of many unknown facts, they are urged as starting points from which to base other and more accurate studies. The medical treatment from this point of view is very suggestive. Obviously the removal of the opium is not the cure. The various methods of removal, detailed with great exactness, as if they would apply to each case, are unfortunate reflections of the failure of the writers, and are based on the assumption that all cases are the same, and the removal of opium is the great essential in the treatment.

Basing the treatment on the clinical study of the case, it

will be evident that where an opium diathesis exists, the withdrawal of opium should be very gradual. The treatment and surroundings should be arranged with great care and exactness. Such persons should live in an institution for a long time, or be under constant medical care. The danger of relapse and the future of such cases will depend entirely on the conditions of life and surroundings. Rapid reduction and any heroic treatment is never permanent, even with the consent of the patient. Specific, faith cures, or any measures that promise speedy cure, are failures from the beginning. The road back to approximate health is straight and narrow, and only along lines of applied science. Where the history of a *neurotic diathesis* is present, the withdrawal of the opium should be equally slow.

More attention must be paid to the brain and nerve nutrition. The removal of opium may be followed by the appearance of very serious disorders, such as epilepsy, hysteria, complex neuralgias and paranoics phases, alcoholism, and various other neuroses. The slow withdrawal of opium enables one to discover and anticipate these neurotic troubles which have been masked before. In one case, suicidal melancholy ; in another, hyperæmia of the brain, with delusions ; in the third, irritation and delirium ; in the fourth, hysterical spasms appeared when the opium was removed. I have seen two cases of general paralysis suddenly spring into great activity after the opium was taken away. This condition was not suspected before. Alcoholism is a very common sequel after the removal of the opium. *Cocaine, chloral,* and almost every drug that has narcotic properties, are also very common entailments.

While these are extreme cases, they are likely to appear at any time. Great care should be exercised in using other narcotics to lessen the irritation from the withdrawal of this drug. Foods and tonics should be given. These cases require the same general treatment as neurasthenia and other states of brain exhaustion. They are drug-takers and will resort to anything for relief. They are secretive, and require more care and more mental remedies, with long, exact hygienic surroundings.

Where the opium addiction has apparently come from bad nutrition and faulty elimination, with auto-intoxication, the treatment is very hopeful. A long preliminary course of baths, mineral waters, and tonics should precede the removal of opium. Then the opium may be removed slowly and without the knowledge of the patient. In proper surroundings, with frequent baths, little danger of relapse or suffering will follow. Careful study and treatment of nutrition and digestion will fully restore the case, and relapse seldom occurs except from failure or neglect of the surroundings.

In the last class, where opium is taken and apparently follows from the entailment of some injury or disease, or the exhaustion of old age, a preliminary period of treatment seems to be required. Often the opium can be abandoned at once for some milder narcotic, and from this, by gradations, discontinued entirely. Full knowledge of the diseased states present will always suggest the lines of treatment. In some cases the opium should not be removed ; its diminution and concealment is required. In others, its rapid removal is essential. Many varied and difficult questions will appear in these cases. The more accurately the diseased states, also predisposing and exciting causes, the diathesis, and varied influences which have caused opium to be used, are studied, the more accurate the treatment. As in many other diseases, the causes may be anticipated, also neutralized and prevented. Opium-taking should be seen as a symptom ; remove or break up the cause, and this symptom disappears.

Routine treatment, either by slow or rapid reduction of the opium, is not wise. The substitution of other narcotics is equally unwise. In a certain number of cases the withdrawal of opium only unmasks more serious diseases, and is often wrong. A case of general paresis is now under treatment for the opium addiction. Before this opium addiction began, the patient caused great distress by his delusions and extravagantly strange conduct. This treatment is wrong. A rheumatic woman of seventy is going through the same course to be free from opium, which has made life tolerable for ten years past. The treatment of opium mania is something more than the application of means and remedies for withdrawal of the drug with the least suffering. The symptomatology and organic lesions often date back to other causes more complex than opium. The treatment must begin by their removal. The general or special diathesis must be treated ; the nutritive disorders, intoxications, and starvations must be recognized and removed. The influence of pathological states from previous disease and injury must be ascertained and treated. The power of environment, climate, occupation, and idiosyncrasies are also powerful factors to be considered.

These are the essential facts and conditions which must enter into the practical treatment. Among the many important problems, that of prevention promises the greatest possibilities. A recognition of the neurotic diathesis and other predisposing causes would enable the physician to successfully guard against its approach.

CHAPTER XXXII.

OPIUM INEBRIETY (CONTINUED).

In the special treatment: 1. We have to deal with an individual whose will-power is subverted. To him the enslaving drug has become as much a necessity of existence as his food and drink. Any treatment which depends upon his own volition must fail. For his own must be substituted the control of another sound will. As a rule, removal from home is essential to secure this control. As in insanity and hysteria, strangers have far more control than relatives or friends. It has the further advantage of breaking up the accustomed train of associations, which is always a great aid in overcoming a confirmed habit. Special asylums have their advantages (if under proper management) and their disadvantages. I shall not discuss this point.

The choice of attendant is of great importance, as upon his or her trustworthiness and efficiency the result may often depend. In the case reported, the firmness and tact of the nurse, her readiness with massage, bath, medicine, or nourishment, etc., enabled the reduction to be made rapidly, and assisted greatly in mitigating the prostration and suffering of the patient. With inflexible will she combined a patience and sympathy which made the patient feel she was a strong friend to help, not a jailor or detective, and was thus a model of what is needed in the attendant.

2. Control of the patient having been secured, how shall the drug be taken from him? Three methods have their advocates: (*a*) immediate and entire withdrawal; (*b*) gradual reduction; (*c*) rapid reduction.

329

Under the first the sufferings are intolerable, the prostration great and dangerous and it does not offer any great security against relapse.

In the majority of instances the rapid reduction is the wiser means between the two extremes. The rapidity should vary with the case, and should be such as not to involve extreme suffering or great prostration.

3. We have to deal in all cases of long standing with an emaciated body and starving nerve centers. At the same time we have complete anorexia and feeble digestion, perhaps nausea and vomiting. The feeding of the patient becomes, therefore, one of the most important, and perhaps most difficult parts of the treatment.

Often it is well to begin with exclusive milk diet (peptonized, if necessary). Systematic feeding of small quantities at frequent intervals is usually best. Confinement to bed during early part of treatment will promote the nutrition. At the same time it reduces to the minimum the tax upon the shattered nervous system. For the same reason, as well as for the sake of preventing the clandestine supply of the drug, seclusion is best until convalescence is well established.

The good results of the "rest treatment," as advocated by Mitchell, *i. e.*, seclusion, confinement to bed, forced feeding, massage, and electricity, with gradual (usually *rapid*) reduction of the drug, are permanent.

4. The use of various mechanical agencies for the relief of pain, quieting the nervous system, inducing sleep and promoting nutrition—massage, electricity (both faradism and galvanism), hot baths, Turkish baths, the cold shower-bath. Dr. Jennings recommends the hammock for the restlessness and desire for constant motion, so often a distressing symptom.

5. Medicinal agents to meet the various indications of each case.

The observations of Drs. Jennings and Ball, of Paris,

upon the sphygmographic tracings of the pulse of habitués we believe, have laid the physiological basis for a rational system of medication.

These observations, which have been confirmed by others, show, "that the pulse of a morphine habitué in a state of privation . . . . caused by want of cardiac impulsion, together with a resistance to the passage of the blood in the vessels. A hypodermic or morphia given at this moment restores the normal state of the circulation. . . . . The study of these tracings suggested the use of cardiac tonics and stimulants as substitutes for the morphia during the progressive reduction." The drugs chosen were : " Sparteine, on account of the facilities it offers for hypodermic injection, and producing thus a rapid and evident effect ; and trinitin, because of its congestive effect on the head and its calorific effect upon the body generally."

Dr. Jennings uses these remedies in the gradual suppression of the drug when the reduction has reached such a degree as to bring on the symptoms of deprivation. It is not to be understood that these drugs take the place of morphia, *i. e.*, that it can be at once omitted without the usual suffering ; they are but aids in mitigating that suffering by counteracting some of the circulatory disturbances upon which it largely depends.

The evidence of clinical experience is largely in favor of heart tonics and stimulants rather than of sedatives. In the case reported, no sedatives were given, yet after the first few nights, sleep was good. The glonoin had certainly a good effect, being given at the time when the symptoms of the craving came on. Quinia was used as a stimulant to the heart and the cerebral circulation. Strychnia was given as a heart tonic after complete withdrawal of morphia.

When nervousness is great, or insomnia does not yield to other means, drugs may be necessary. In these instances, cannabis indica in large doses (½ to 1 drachm of fl. ext.), sulfonal, chloralamid or bromides will often render good service. Chloral is used by Erlenmeyer, condemned by Aurleck and others. Dr. Jennings seems latterly to have usually substituted digitalis *per os* for spartein hypodermatically. Quinia has seemed to me in many cases of distinct value. Strychnia is one of the best heart tonics in the pharmacopœia.

Obersteiner is almost the only writer of note who now speaks well of cocaine. If used at all, it should never, of course, be placed in the hands of the patient himself. The fluid extract of coca has been highly spoken of by several writers to relieve restlessness and depression. Valerianate of ammonia has been a common favorite since the time of De Quincy.

This method, which, so far as we are aware, is original with and peculiar to Dr. Mathison, is merely a new application of a well-established principle, for the power of the bromides to subdue abnormal reflex irritability is so constant that it may be looked upon as an invariable sequel of such medication. Dr. Ed. H. Clarke, in his valuable treatise on the bromides, says : " Diminished reflex sensibility, however different physiologists may explain the fact, is one of the most frequent phenomena of bromidal medication that has been clinically observed, and is, therapeutically, one of the most important." The testimony of other distinguished observers is to the same effect—Gubler, Guttman, Laborde, Voison, Damourette, Eulenberg, Claude Bernard, Brown-Sequard, Echeverria, and Hammond all giving evidence as to the power of these agents to impair the control of the spinal cord over reflex manifestations, and, at the same time, exert a marked influence over the general nervous system. Admitting that the symptomatology of opiate abandonment pertains almost exclusively to the functions over which the bromides exert so decided a control, we have in the treatment of opium inebriety a new field presented for the exercise of this valuable property, and the fact, proven conclusively by our experience, that it *does* exert this happy effect, fully supports the idea advanced as to pathology of this disease.

In speaking of the bromide of sodium, let it be understood that we refer entirely to the influence of the *continued dose*, by which we mean its administration three times in the twenty-four hours, at regular intervals, so as to keep the blood constantly charged with the drug. A most important difference, physiological and therapeutical, exists between the effect of this mode of exhibition and that of the single dose, or two or three doses so nearly together as to form practically one. for in the former case the system is constantly under the bromide influence, while in the other, the drug being largely eliminated in a few hours, the blood is nearly free from it a large portion of the time. Results obtainable from the continued use cannot be gotten from the single dose, and, as a consequence, its value is far greater in the disease under consideration.

Again, the action of the continued dose being somewhat remote, three to five days usually elapsing before there is decided evidence in this direction, much more desirable results are secured by its employment for several days *prior* to an entire opium abandonment, meanwhile gradually reducing the opiate, than if the withdrawal be complete and then reliance placed on the bromide to control the resultant irritability ; for, in one instance, the maximum sedative effect is reached at the period of maximum disturbance from the opium removal, and its counteracting and controlling effect is far in excess of that to be had from its employment subsequent to the lighting up of the nervous irritation. What, then, we style *preliminary sedation* forms a peculiar and most valuable feature in our administration of the bromide, and it is this particular point we commend to you, our experience having convinced us we have in it an unequaled means of obviating the suffering incident to the treatment of this disorder.

The value of the various bromides depends on their proportion of bromine. Bromide of potassium contains 66 per cent., sodium 78, and lithium 92 per cent. We should, therefore, expect a more powerful influence from the latter agent, and, according to Wier Mitchell, it has a more rapid and intense effect. Bromide of sodium being richer in bromine, and pleasanter to the taste, we prefer it to potassium.

Either of the bromides, in powder or concentrated solution, is somewhat irritant, sometimes provoking emesis, and, in any event, delaying its absorption. A practical point, then, is that it be given largely diluted. Dr. Clarke says, "There should be at least a drachm of water to each grain of the salt." We give each dose of the sodium in six or eight ounces of cold water, and have never known it to cause vomiting.

Another important feature relates to the time of its employment. We usually administer it at 10 A. M., 4 and 10 P. M., or half an hour before each meal. Given thus, largely diluted, it is probably absorbed in half an hour, and the effect of the continued dose rapidly secured.

To produce the requisite degree of sedation within a limited period, it is essential that the bromide be given in full doses. I am convinced that failure in its use, in any neurosis, is very often due to a non-observance of this point. Our initial dose of the sodium is 30 grains, twice daily, time and mode as stated, increasing the daily amount 30 grains each day, i. e., 40, 50, 60 grains and continuing it eight days, reaching a maximum dose of 100 grains twice times in the twenty-four hours. It is then discontinued, but its sedative effect persists and somewhat increases for two or three days following, and then begins to decline. This period may last a week or more depending on some state of the blood and power of elimination. During this week of bromidal medication, the usual opiate is gradually reduced, so that on the seventh or eighth day it is entirely abandoned. A decrease of one-third or one-half the accustomed daily quantity is made at the outset, experience having shown that habitués are almost always using an amount in excess of their actual need, and this decided reduction occasions little or no inconvenience. Subsequently, the opiate withdrawal is more or less rapid according to the increasing sedation, the object being to meet and overcome the rising nervous disturbance by the growing effect of the sedative—in other words, maximum sedation at time of maximum irritation.

Having secured the sedative effect desired, the object is to eliminate the bromide as rapidly as possible, and as the skin· and kidneys form the only outlets, recourse is had at once to diaphoretics and diuretics. Of the former, hot and steam baths are to be relied upon. And, of the latter, digitalis, in infusion, or, if bulk be objectionable, Squibb's fluid extract combined with potass. acet. and spirits æther nitrosi. The bromide itself increases renal secretion, and, aided by the others, it passes from the system in a few days. The bromide and opiate having been discontinued, restlessness, more or less prominent, from 20 to 56 hours, invariably supervenes. It is controlled by codeine, 1 to 3 grains subcutaneously, or by mouth, every 2 to 4 hours, and this is continued, decreasing the dose or increasing the interval till no longer needed, but is greatly relieved by hot—not warm—baths, temperature 110° to 112°, 15 to 30 minutes duration, repeated as required. They are often signally effective. We have known a patient fall asleep, snoring vigorously, while in a bath.

Sleeplessness is always more or less prominent after opium abandonment. During the first six nights, sulfonal or trional in 30 or 40 grain doses is given. Afterward, such hypnotic as seems best suited to the case ; chloral is most effective, in 20 or 30 grain doses at bedtime. Often smaller doses may be given at intervals of two hours until sleep follows, with good results.

A peculiarity of this insomnia is, that it is most marked in the early morning. Slumber comes readily enough at night, but patient awakes at two, three, or four oclock, and finds further sleep impossible. Often it is well to defer the sleeping draught until this time. This waking tendency gradually diminishes, and ultimately disappears.

Chloral, given during the early opium abstinence, has, with us, not acted kindly as a hypnotic, but produced a peculiar intoxication, though we have never noted the wild, maniacal delirium mentioned by Dr. Levenstien as occurring during this period in his cases. As soon as possible it should be discontinued, and sleep secured by a fatiguing walk, a half hour's warm bath, a light lunch, or glass of milk—one or all, before retiring.

For three or four days following the opiate withdrawal, the diet should be *exclusively* of milk combined with lime-water—one or two ounces, with one or two drachms respectively—every hour or two. It is very seldom rejected, and is preferable to beef tea, or anything else. Afterwards, a full, solid diet may be resumed, soon as practicable.

While diarrhœa is the decided *exception* under this plan of treatment, we still deem it best to keep the bowels in good condition, and administer the first night, a mercurial cathartic sufficient for several full evacuations, followed, during the bromide giving, by daily laxative enemas, or doses of Hunyadi water.

Debility, of varying degree, due to the opium abstinence, and bromide relaxation, is among the sequelæ. It decreases with the increasing bromine elimination, aided most effectively by general faradization, twenty minutes morning seances daily, after the restlessness subsides, and strychnia 1-20 gr. thrice daily, combined with iron, quinine, phosphorus, digitalis, or cod liver oil, as most required.

The following formulæ are valuable :

℞. Strychnine, two grains; Muriated Tinct. Iron, five ounces; Tinct. Digitalis, and Glycerine, of each, two and one-half ounces. M. Dose—One to two drachms, three times daily.

℞. Strychnine, four grains; Dialized iron, five ounces. M. Dose, one-half drachm, ter in die.

℞. Strychnine, two grains; Dilute Phosphoric acid, and Syrup of Ginger, of each, two and one-half ounces. M. Does—One drachm thrice daily.

℞. Pyrophosphate of Iron, 5 to 10 grs., Quinine, 2 grs. at a dose. M. Pill, or solution, three times a day, if the appetite be slow in returning.

℞. Comp. Tinct. Quassia, one drachm; Tinct. Capsicum, ten drops. M. For one dose, diluted, 20 minutes before each meal.

Strychnia is not advisable *during* the bromide adminis-tration. Being decidedly antagonistic—one causing relax-ation and deficient reflex excitability, the other, just the reverse—the desired sedative effect may be materially delayed if they be given together. Subsequently, it is the most valued general tonic at command, and may be con-tinued in varied combination for weeks. With the strictly medicinal course are to be employed a full nutritious diet, out-of-door exercise, especially walking, and varied social enjoyments—in fact, anything that can exert a roborant effect on mind or body.

Surprise may be expressed, and objection made regard-ing the extent of the bromide doses, but the fact must never be overlooked that we are not to be governed in the giving of any remedy by the mere numerical amount of drops or grains, but by the effect produced. Again, I am led to think that one effect of opium addiction is a peculiar non-suscep-tibility to the action of various nervines, necessitating their more robust exhibition to secure a decided result. More,. and most important of all, under the influence of certain

abnormal conditions, doses which ordinarily are toxic become simply therapeutic. The annals of medical literature
abound with illustrations in support of this statement, and
among the most striking may be noted the following : Dr.
Southey read before the Clinical Society of London notes of
a case of idiopathic tetanus, which occurred in a boy ten
years old. The first symptoms of trismus were observed
two days after a severe fright and drenching due to the upset of a water-butt. They steadily increased up to the date
of his admission to St. Bartholomew's hospital, upon the
eighth day of his illness, when the paroxysms of general
opisthotonos seized him at intervals of nearly every three
minutes. Each attack lasted from fifteen to thirty seconds,
and although between the seizures the muscles of the trunk
became less rigid, those of the neck and jaw were maintained
in constant tonic cramp. The patient was treated at first
with chloral, ten grains, and bromide of potassium twenty
grains, every two hours, and, afterwards, with the bromide
alone in sixty grain doses every hour and a-half. When
about two ounces were taken in the twenty-four hours, the
attacks became less frequent, but at first each separate seizure was rather more severe, and upon the evening of the
eleventh day, he was able to open his mouth better.

On the thirteenth day the bromide was decreased to
twenty grains every three hours, and on the fourteenth day,
was discontinued altogether. When the bromide had been
omitted for twenty-four hours, the attacks returned at intervals of an hour, and the permanent rigidity of the muscles
of the neck was re-established. His condition now steadily
became worse, so that on the eighteenth day of his illness
it became necessary to resort to the previous large doses—
one drachm—every hour and a-half. After three such doses,
the expression became more natural, and he was able to
open his mouth again ; but it was not till the twenty-fifth
day of the disease that it was possible to discontinue the
remedy. The patient remained in a state of remarkable

prostration and drowsiness, sleeping the twenty-four hours round, and only waking up to take his nourishment for eight days, and passed all his evacuations under him. He subsequently steadily and rapidly convalesced. The bromide produced no acne or other disagreeable symptoms, and certainly appeared to exert marked inhibitory influence upon the tetanus.

Surely, under ordinary circumstances, no one would think of giving such extensive doses of the bromide, but here, under the antagonizing influence of the intense reflex irritability, their effect was vastly beneficial, conducing, unquestionably, to the patient's cure.

Given, as we recommend, no effect is usually produced by the bromide before the third day. From the third to the fifth, an unpleasant taste is complained of ; the bromic breath begins ; the patient is disposed to drowse, and there is a growing indisposition to muscular exertion. From the fifth to the seventh, these symptoms increase—the tongue begins to furr ; the odorous breath is marked ; the drowsiness deepens into sound sleep, more or less prolonged, and the inaptitude for physical exercise becomes so decided, that patients generally take to bed on the last day. The following two or three days—during the period of maximum disturbance from the opium withdrawal—are characterized by a persistence of the symptoms alluded to. Patient remains, more or less restlessly, in bed; general relaxation is decided; the pulse is less frequent—usually about 60 ; the voice somewhat weakened ; pupils dilated ; the renal secretion augmented—though, sometimes, diminished ; the saliva increased and rather viscid, and mild hallucinations of sight and sound—almost always of sight—occur, occasionally, for three or four days, accompanied with a peculiar aphasic tendency, as shown by substituting one word for another— Mediterranean for Mississippi; Brown, instead of Iowa, etc. This curious symptom may occur at increasing intervals for several days.

Dr. Clarke refers to such instances. He says: " They
are hints of a distinct organ of language, and suggest the
notion that, inasmuch as the drug we are considering paralyzes reflex, before it does general sensibility, language
may be the expression or correlation of a peculiar reflex
power."

After the ninth or tenth day, the bromidial manifestations gradually disappear, so that within two weeks from
beginning of treatment, patient is generally up and the
only prominent symptoms remaining are the debility and
insomnia. Tonics, hypnotics, and vis medicatrix naturæ
effect speedy convalescence, and—where treatment is begun
on entrance—patients are usually dismissed, cured, within
a month.

More than one week's employment of the sodium is not
advisable lest the hallucinations become unpleasantly persistent ; and cases will present in which a minor degree of
administration—five or six days—will suffice.

Marked general debility contra-indicates the bromide,
and a tonic course should precede it.

Granted a case suitable for treatment, this method may
be summarized as follows : Opiate reduced, at once, to
one-half or two-thirds usual quantity. Subsequent gradual decrease and entire withdrawal in seven or eight
days. Mercurial cathartic, first night, followed by daily
laxative enemas, or Hunyadi water. Bromide of sodium,
30 grain doses, increased 30 grains daily, in six or eight
ounces water, on empty stomach, continued 5 to 7 days.
Restlessness following opium abandonment met by hot
baths, 100° to 110°, ten to thirty minutes each, often as
required. Bromide eliminated by diuretics—digitalis and
nitre, and diaphoretics—hot and steam baths. Insomnia
relieved by chloral, combined, if need be, with Indian
hemp or hyoscyamus. Diet exclusively milk and lime-
water first three days of opium abstinence. Full diet
resumed soon as possible. Debility removed by generous

living, general faradization, strychnine, iron, quinine, etc., with out-of-door exercise and varied social enjoyment.

For relief of neuralgic pain varied measures suffice. Leading the list are electricity and the local use of ether. As to the value of the galvanic current in migraine and other neuralgias, so common in opium habitués, and the manner of using it, the reader is referred to papers by the writer. The same agent is effective in relieving limb and lumbar pains, though here a much stronger current is required than can be used with safety about the head. Sometimes a faradic current acts well, and when one fails, trial should always be made with the other. Local hot baths are often of great service.

Regarding the ether, those who have never used it will, we think, be surprised at its pain-easing power. In either way applied—spray, drop, or lavement—it is potent for good.

These three—electricity, ether, hot water, are valued anodynes, and one special point in their favor is entire freedom from unpleasant gastric, or other result.

Other remedies relieve, at times, of the coal-tar salts, phenacetine, or phenocoll, 10 to 15 gr. doses are best. It has often a hypnotic effect. A valued external anodyne is menthol, 1 part ; chloroform, 10 parts ; ether, 15 parts ; used as spray.

Under this plan of treatment marked disorder of stomach or bowels is rare. Our rule is to give a mercurial or other cathartic at the outset, if there be alvine torpor, and then secure regular action by such laxative as seems best. If restraint be needed, large enemas of hot water may be used. This failing, 1 to 3 grs. sulpho-carbolate of zinc, 10 to 20 minim doses of fluid extract of coto, in capsules, or 40 to 60 grains of subnitrate of bismuth every four hours. If, however, it persists, the best thing is a full opiate—tinct. opii, per mouth or rectum, at bedtime, preferred. This promptly controls, gives a full night's sleep, and the trouble seldom returns. Fear of a bad effect on convalescence is unfounded.

Diet is not restricted, unless the condition of stomach or bowels demands. We have again and again seen patients recover who did not vomit once, or who had only two, three or four movements daily. The excessive vomiting mentioned by Levinstein and Obersteiner—abrupt disuse—we have never noted. The former thought the collapse—which we have never seen—in several of his cases was due to vomiting and purging. More likely the largest factor in causing it was the exhausting mental and physical suffering which his method entails. If the stomach rebels, entire rest for a time, or milk and lime-water, ale and beef, malted milk or bovinine, in small amount, may act well. If not, sinapisms, ether, faradism, or chloroform, alcohol and ice are of value. All failing, a full opiate hypodermic will promptly suffice.

Twenty-four hours after the opiate-quitting, patients are directed to bed, and kept there two to four days, for we are convinced that rest is an aid of great value. Erlenmeyer says, "The best remedy is rest in bed. The importance of quiet, rest in bed, and warmth in promoting restoration during the abstinence struggle, cannot be overestimated. I order every patient to bed at the start, and can state with confidence that those who submit to this till I allow a change will get along more easily and satisfactorily during the treatment than others who do not obey, but who insist on moving about or having the run of the premises."

Having thus crossed the opiate Rubicon, treatment pertains, mainly, to the debility and insomnia. For the former, coca leads the list. Of fluid extract, 2 to 4 drachms, or cocaine, 1 to 2 grains, with other tonics, 3 or 4 times daily, decreasing as need lessons. As a rule, its use is ended in a fortnight. To remove the mental and physical depression, the minor neuralgias, and the desire for stimulants sometimes noted nothing equals it, and full doses of tincture of capsicum often add to its value.

Another agent of much service is general faradization, twenty minute *séances* daily. This imparts a feeling of exhilarating comfort, but care must be taken not to overdo, for a current too strong or long makes mischief, overstimulating and exhausting to the extent, it may be, of several days' discomfort, which nothing but time will remove.

Faradism also acts kindly in easing the peculiar unrest —"fidgets"—and the nagging aches in legs during convalescence. It may be applied in the usual way, or through the special electrodes we have devised.

Galvanism is another general tonic of value. Our method is positive pole to nape of neck and negative to epigastrium for five minutes, then the former behind the angle of each jaw for a minute or two, making entire *séances* seven to nine minutes.

Another valued tonic is the cold shower-bath. With many it is a great invigorator, and patients who dread it, at first, came to appreciate it highly.

Internal tonics have a place in the roborant regime. Most habitués are below par, and it is our custom to give such, from the start—phosphorus, strychnine, arsenic and quinine, combined. After the opiate-quitting, coca, in some form, can be added. If anæmic, ferri tincture, or Blancard's pills. Caffein is of value. It is stimulant, tonic and diuretic. We sometimes give it with codeine and cocaine. Digitalis is often useful. In some cases, cod-liver oil is of service—with pepsine and quinine, with malt, with phosphates, or plain—and may be given for months.

Some anorexia is usually present, yet it may not prevent the regular meal, and need never occasion anxiety, for it will likely give place to a vigorous appetite, which may be encouraged to fullest feeding short of digestive disaster. If it be slow in returning, half grain doses of cannabis, an hour before meals, often have a marked action.

Regarding the insomnia, Levinstein said : " Sleeplessness, which is generally protracted up into the fourth week,
is very distressing." Our record differs. Wakefulness is
an invariable sequel, but usually not so marked nor prolonged, and in ordinary cases recovery can generally be
promised, without the loss of a single entire night's sleep.
We have known a patient able to dispense with hypnotics
in five, others in eight, and the average, in a series of cases,
was eleven nights.

The insomnia is of two kinds. Most patients secure
sleep on retiring, but waken early—3 or 4 o'clock—and fail
to get more. Others remain awake nearly all night before
slumber comes, and these usually require soporifics the
longer. For relief of this, cannabis indica will often suffice.
The hemp is given in 40- to 60-minim doses, in capsules, or
mixed with glycerin, or ginger syrup, two hours before bed-
time. There may be noted, in some, laughing and talking
during the first hour, tending to sleep in the second.
Many require nothing else. At the end of a week it is lessened, and usually, ended in ten or twelve days.

Other hypnotics—chloral, chloralamid, trional, sulfonal,
paraldehyde, hypnal—in full doses, often work well.

Chloral, during the first 5 or 6 nights of opium abstinence, fails as a soporific, often causing a peculiar excitement or intoxication—patients talking, getting out of bed
and wandering about the room—followed after several
hours, by partial sleep. Later, in full doses—we prefer
40 grains at once, rather than two 20-grain doses—alone, or
with a bromide, it can be relied on.

If, as rarely happens, the sleepless state is so pronounced
or prolonged as to distress patient, we never hesitate to
give a full opiate, by mouth, and with good result. Erlenmeyer says : " In such cases there remains nothing to do
but to resort to morphine. I give, then, the alkaloid inter-
nally on two consecutive evenings ; a certain cumulative
effect takes place. The first night in the dose of $\frac{1}{2}$ grain,

there is usually no sleep; but on the second night, after giving the same dose, a sound sleep of six hours' will ensue. I have not observed any special danger from these resumed doses of morphine, although I feared it; but after I was constrained in several bad cases, when every other medicine had failed, to resort to this, I was convinced that my fear was groundless."

In all cases drugs should be dropped as soon as possible, and sleep secured by a walk or other exercise—an electric *séance*, a Turkish or half-hour's warm bath, a light meal, a glass or two of hot milk, one or more of these before retiring. Patients, whose slumber ends early, often note a peculiar depression on waking, and if so, a lunch, hot milk, cocoa, coffee, beef, or bovinine should be at command.

It may be well in passing to refer to certain minor sequelæ and their treatment. If dyspnœa or palpitation, a stimulant—coca with capsicum or Hoffman's anodyne with aromatic spirits of ammonia—will promptly control.

If aching pains in the calves, strong galvanic or faradic currents, hot water, massages, or ether will relieve. If a peculiar burning in soles, mustardized foot-baths. If marked hysteria, ether inhalations.

Belly pain may be eased by hot fomentations, or full doses of ether in hot water, or camphor with capsicum. The latter, with atropine injection, act happily in ovarian irritation.

Very seldom unrest and insomnia compel hyoscine. If so, hydrobromate 1-100 to 1-20 grain, hypodermically, or, double, by mouth.

CHAPTER XXXIII.

ETHER INEBRIETY—ITS HISTORY AND PROGRESS.

Ireland has in the nineteenth century presented to the world two interesting and remarkable series of inebrio-psychological phenomena. In 1838, a simple-minded Roman Catholic priest, Father Mathew, adopted and began to advocate the practice of abstinence from all intoxicating drinks. So amazing was the impression made by him that, in three years, the roll of the teetotal pledges which he had administered exceeded 5,000,000, in Ireland, in addition to large numbers in England, Scotland and America. The reality of this epidemic of temperance was attested by the statement of the Chief Secretary, in 1840, that "the duties of the military and police in Ireland are now almost entirely confined to keeping the ground clear for the operation of Father Mathew."

Though this great wave of sobriety has gradually receded, till now the extent of drinking in Erin is simply terrible, I am every now and again meeting professionally with sons and daughters of Hibernia, who glory in their steadfastness to the pledge which they so long ago took at the hands of the Irish apostle of temperance. An accurate study of this unique crusade would, in psychological results, amply repay the labors of any earnest student of mental science.

Curious to relate, the other series of inebrio-psychological phenomena is an experience in an opposite direction—an experiment, so to speak, not, as in the former case, in temperance, but in intemperance.

347

The disease of inebriety or narcomania (a mania for intoxication by any kind of narcotic or anæsthetic), may besides other phases, assume a form correspondent to the particular inebriating substance. It may, therefore, be interesting to glance at the origin and growth of this new mode of inebriate indulgence, as this is the first opportunity afforded to us of observing the rise and progress of such a process in a community.

The centre from which ether drinking spread was the town of Draperstown (with a population of some 300), in the southern part of the County of Londonderry. Before Father Mathew's abstinence propaganda, ether drinking was there unknown. Between 1842 and 1845 a local medical practitioner, in response to a request from a few newly-pledged abstaining converts for something the taking of which would not violate their vow, gave them a drachm of ether in water. So far as I can ascertain, this was the *fons et origo mali*. A desire for more frequent doses grew upon the ether drinkers, and the practice spread in and around Draperstown till there was a shop for the sale of ether, in one town, to every twenty-three of the population. In the session of 1855–6, an Act was passed by the British Legislature allowing spirits of wine to be used duty-free in arts and manufactures, provided it was made nasty as a drink (which the Government, in their innocence, supposed would prevent people from drinking it) by the addition of a minimum 1-9 of methylated spirit. As ether prepared in this way is much cheaper than ordinary sulphuric ether, this cheap production of "methylated ether" caused the consumption to increase "by leaps and bounds."

The present ether area was, from its mountainous features, a central locality for the illicit distillation of whiskey. Owing to the activity of the police and the making of roads, this illicit traffic was effectually stamped out. The disappointed cheap whiskey drinkers found a cheap unintoxicant in ether. Mr. H. N. Draper first called attention to Irish ether drinking in 1877, followed by Dr. B. W. Richardson, about 1879, and by Mr. Ernest Hart in 1890.

Ether drinking was in a year or so gradually introduced from Draperstown into the neighboring town of Maghera, and soon extended its sway till it occupied an area of somewhere about 295 square miles, with a population of nearly 79,000 souls. This area may, in general terms, be said to comprise the mountainous districts, especially of Derry and Tyrone, and to some extent, of Armagh and Antrim. Cases of ether intoxication have occurred in Dublin and other parts of Ireland, in Glasgow (Scotland), in Lincolnshire in England, and I have seen several in London.

All the cases which I have seen in England have been persons of education and refinement, who had first been alcholic inebriates and gradually developed into devotees of these twin poisons. Nearly all of these English cases have been females, the only males having been members of the medical profession. In Ireland, women assert the equality of the sexes by taking their fair share of this form of intemperance. Small farmers and agricultural laborers make up the bulk of the Irish ether tipplers. Workmen, too, are well to the front. But the practice is by no means confined to these classes. Members of the learned professions have their representatives. Etherists are to be found at almost all ages from puberty onwards. Sturdy Irish lads and beautiful Irish lasses, brimful of Hibernian wit, as well as "60-year olds" of both sexes, are slaves to ether drunkenness. The mother may be seen with her daughters, and maybe a neighboring Irishwoman or two, at a friendly ether " bee." The habit has become so general that small shopkeepers treat the children who have been sent to purchase some article, with a small dose of ether, and schoolmasters have detected ether on the breaths of children from 10 to 14 (or even younger), on their arrival at school.

Some critics have endeavored to lay the blame of this new development of inebriety on the Roman Catholic

ıeligion. Nothing could be more unwarrantable and
unfair. The disease has spread principally among Roman
Catholics, simply because this is the creed of the greater
part of the population. One Protestant village, Tobermore
is as bad as any other place. All my cases have been Prot-
estants.

The amount swallowed at a draught varies mainly with
the stage of education in ether consumption. A novice
will find a drachm (a teaspoonful) sufficient. Gradually
the wished-for effect demands an increased dose, till ¼ of
an oz. may in time become the ordinary "peg" of an
accomplished drinker, to use the phraseology of Anglo-
Indians. These are average quantities of a so-called
"moderate" drinker. More "seasoned casks" have a
higher capacity, many toping off a half a wineglassful as
unconcernedly as an average Englishman would drink a
glass of claret, or an average American a glass of champagne.

The amount of ether consumed in a day is often remarkable. A confirmed ether inebriate will take a much larger
dose than any I have just enumerated, and repeat the dose
three, four, five, or even six times in the twenty-four hours,
when "on the spree." Indeed, in some cases, half a pint
has been the regular daily allowance of constant (or habitual) inebriates. In England I have known an ether inebriate use a pint of ether by inhalation every day. In Ireland,
many persons keep themselves intoxicated pretty well during the day for the sum of sixpence—taking two penny-
worth at 10 o'clock, 1 o'clock and 4 o'clock. What a para-
dise for drunkards ! Drunk three times a day for 13
cents !

In England, in my own practice, the majority of ether
drinkers have inhaled the poison. In Ireland the universal
method is drinking. By the latter mode the ether is taken
"neat." Owing to an idea that ether, like whiskey or
brandy, should be drunk diluted with water to sheathe the

virulence of the poison, the uninitiated and ignorant Englishman, when in Ireland, sometimes mixes his ether " peg" with water, "just to try the stuff, you know." Ludicrous failure awaits him, for, unlike ardent spirits, ether is but sparingly soluble in water. The pungency of ether, except to those who have " finished their education," generally calls for an "overture" to the "act" of ether swallowing. Scene I.—The mouth is washed out with cold water. Scene II.—A draught of cold water is drunk. Scene III.— The ether is swallowed " neat." Scene IV.—The performance closes with a second and final drink of cold water.

The preliminary draughts of water are to cool the mouth and throat, and the post-ether draught is " to keep the ether from rising." The washing of the mouth is soon omitted. By-and-by the preliminary draught of water follows the same fate, the ether dose and the succeeding draught of water being the commonest method. As his education advances, the etherist dispenses with water altogether. He may for a while, especially when drinking an unusually large dose, hold his nose with one hand, but probably ends by despising all precautionary safeguards, and by simply drinking his mouthful of ether at a gulp.

Ether purus of the British Pharmacopœia was at one time affected by my inebriate patients. This pure ether $(C_2H_5)_2O$, which is free from alcohol and water, has been in my hands the only ether preparation which has proved to be without complicatory drawbacks when used as an anæsthetic (Brit. Pharmacopœia, 720 ; U. S., 725). *Ether* of the B. P. (sulphuric ether) was, however, the article generally used for purposes of intoxication. It contains 8 per cent. of alcohol and water with 92 per cent. of *ether purus*, and is soluble in all proportions with rectified spirit, but in only 1 in 10 with water. The specific gravity should be (B. P.) .735 ; (U. S.) .750. It is a swift, potent, diffusible stimulant, narcotic, anæsthetic and antispasmodic, of great value in medicine. It has a strong, penetrating odor, is sweetish, hot, burning and pungent to the palate.

To America the whole world owes a deep debt of gratitude for the introduction of ether as an anæsthetic by Dr. Morton, in Boston, in 1846, and any saddening misuse of this grand mode of alleviating human suffering ought not to lessen our appreciation of this splendid boon to humanity.

By the Act 18 and 19 Vict., the use of spirit of wine, free of duty, was permitted in the arts and manufactures, on the addition of a minimum of 1-9 of wood-naphtha (methylic alcohol or spirit from the destructive distillation of wood, after rectification ; specific gravity .803 B. P.) with a view to prevent this fouled liquid from being drunk as a beverage. Éther prepared from this fouled duty-free spirit is, of course, much cheaper than ether prepared from spirit of wine on which duty has to be paid. The intention, however, was defeated, inasmuch as in the process of manufacture of ether from the fouled spirit, the fouling ingredients (*i. e.*, the methyl products) are destroyed. Thus, contrary to the general belief in what is commonly called "methylated ether" being as nasty as the methylated spirit which is used for lamps and for polishing purposes, "methylated ether" is to the taste hardly discernible from pure sulphuric ether. This so-called "methylated ether" is practically undistinguishable from *ether* (B. P.) at the specific gravity of .717, *i. e.*, when purified. At any other specific gravity, an odor is given off after evaporation. Practically, one cannot discriminate between the ethylic and methylic productions.

Price.—The ether thus prepared from the duty-free spirit (sp. vin. rect. *cum* methylic alcohol), can be produced at as low as one-seventh of the cost of ether prepared from the duty-paid spirit, the latter being bought wholesale at $1.25 per lb., and the former at as low as 16 cents.

Ether is imported mostly from England, partly from Scotland, by larger chemists and druggists in the principal towns of Cookstown, Magherafelt, and Maghera. The large dealers supply small shopkeepers, and also cottagers, who sell in "draughts" (rather less than 2 teaspoonfuls) for one penny. The small shopkeepers also supply the hawkers (who are very often women), who attend fairs and other festive gatherings to dispense the "draughts" of the liquid poison. These "draughts" are also to be had from the surgeries of some medical practitioners, and in cottages or ether shebeen, where the cottager keeps a pig or two, and sells ether, the country people frequently giving potatoes, meal, or other produce in exchange. The hawkers carry about a bottle of ether, and do not scruple at selling to any one, however young, bartering a little for one or two eggs. In this way the children may procure the ether on their way to school.

Intoxication by ether presents one distinguishing feature as compared with alcoholic intoxication. The phenomena are practically alike, but in rapidity of manifestation, alcohol is " nowhere." Indeed, in this respect, ether beats the record. There is the exhilarative stage of morbid exalta- tion, when the fun and exuberant merriment, the latent and ineradicable impulse of one "spoiling for a fight" of the genuine Irishman stands revealed in the twinkle of the eye, and the flourish of the shillelagh. The pleasing but quickly vanishing whirl of enjoyment is followed by an evanescent episode of brain disturbance and mental riot, with muscular paralysis and incoördination. To these succeeds the concluding comatose stage, when the patient is said to be "dead drunk." The shortest period in which I have seen this inebriate panorama move on till it swung round to recovered sobriety has, with alcohol, been six hours. With ether, I have witnessed the entire revolution in less than two hours.

In my observation an alcoholic inebriate career, from start to a fatal finish, has in America been, on an average, one-third of the duration of a corresponding career in Britain. So, curious to say, has the length of an ether

intoxicative paroxysm been one-third the length of an
alcoholic intoxicative paroxysm.   Thus the etherist can
have three thorough "drunks" for one of the alcoholist.
Herein, in addition to the greater cheapness, lies the superior claim of ether to the "greedy for intoxication," the true
"narcomaniac."

Intoxication by ether may be described as "hysterical,"
and intoxication by ether *cum* alcohol as "maniacal."   A
man arrested while drunk on ether alone, would probably
be quite sober by the time the constable had him at the
police station, which might be very awkward for the constable, though the arrest had been made when the man was
in a frenzy of boisterous excitement.   Several deaths from
ether, and ether *cum* alcohol, have occurred.

Little is known of the pathology of ether.   The habit
has been too young to afford opportunities of much post-
mortem examination of ether inebriates.   Premature old age,
an antedated shriveling up of the living frame, attests the
poisonous influence of the destroying agent.   Gastritis
(acute and chronic), debility, dyspeptic distress, epigastric
pain, pallors, tremors, timidity, moroseness, suspicion, nervous prostration, chilliness, a cyanosed or lemon skin, and
an intermitting heart-beat, with exaggerated reflexes, are
prominent symptoms.   I have one such victim in my mind's
eye now.   Fawning, cunning, terror-stricken, this wretched
medical colleague is the incarnation of utter misery.   Not
yet 40 years of age, he shuffles about like a worn-out old
man of 90 after a wasted and mis-spent life.   It has been
urged by some medical authorities that ether is guiltless of
producing any pathological lesion, from the almost lightning rapidity with which its inebriating manifestations
appear and fade away, and from no serious morbid after-
death appearances having been observed.

This conclusion is, in my judgment, premature.   Judging from the symptoms from which I have seen ether
inebriates suffer, I have not the slightest doubt that ether

has a pathological influence on various organs and tissues, and that, if ether drinking could boast of as venerable an antiquity as alcohol drinking, unmistakable lesions would have been but too manifest. What are the forty years of ether consumption by a hundred thousand persons, to the thousands of years of alcohol consumption by at least as many millions of human beings?

Happily, this new form of inebriation is but in its infancy, so there is some hope that its growth may be "nipped in the bud." As, in the conversion of methylated spirit into ether, the nauseous methyl products are destroyed, something might be done towards making the liquor loathsome to the palate by the compulsory addition of the wood spirit *after* the completion of the etherification before the sale of the liquid. This, however, would be but a palliative, for I have had patients under my care who drank methylated spirit (some even from jars with anatomical preparations) and in Edinburgh and Glasgow, Sunday drinking of this nasty beverage recently flourished apace. In a certain locality in the north of Ireland, the drinking of methylated spirit was introduced seven or eight years ago. At first confined to the very poor in a hilly district, it has spread rapidly, till now farm laborers and farmers are daily indulging in it.

There are many inebriates who hate and abhor the taste of the intoxicant which, in their narcomaniac madness, they would barter their salvation to procure.

Another remedy would be the abolition of the retail sale. This would help by putting difficulties in the way of the drinker, but would only mitigate the mischief. Still more effectual would be the scheduling of ether as a poison, the sale of which is restricted to druggists under certain safeguards. This course was so readily adopted for Ireland by the British government in January last, that I have yet hope the day will come when the more deadly allied poison—alcohol—will be placed in the same category, and so dangerous a drug will be relegated to the shelf of the apothecary, its sale hedged in with as stringent precautions as is now the sale of arsenic or prussic acid. But this halcyon era of prohibition will only be attained after a prolonged struggle, amid the howls and groans of an enraged liquordom, whose indignation is concentrated on all who attempt to "rob a poor man of his beer."

It is too soon yet to foretell the ultimate result of the bold step taken by our Government in scheduling ether as a poison, but it has made the procuring of ether for drinking purposes so difficult that for the present the sale has diminished by at least 75 per cent. I fear, however, that the cupidity of some wholesale dealers will incite them to risk the penalties of the law by surreptitious sales, which will speedily be ferreted out by the marvelous cunning of the diseased and demoralized inebriate.

The lines of sound treatment of ether inebriety, and of its prevention by law, must alike be based on an intelligent appreciation of the true character and etiology of ether drunkenness. This is, in reality, but a new manifestation of an underlying morbid condition which renders certain of the sons and daughters of men peculiarly liable to plunge into intoxication. We can never hope to succeed in the cure and prevention of any disease, until we first recognize the presence of the disease itself. The malady of narcomania, as subtle as it is far-reaching in its influence on body, brain and mind, and morals, is a legitimate' outcome of natural law, and we will not be adequately equipped for the fight till we are thoroughly conversant with the laws under which every form of the disease of inebriety is developed and propagated.

CHAPTER XXXIV.

COCAINE INEBRIETY.

The dangers to be apprehended from the abuse of cocaine are probably hardly yet quite realized, at least in this country. A great deal of harm has undoubtedly been done of recent years by the use of cocaine as a help to break off the morphia habit. An exaggerated estimate of the assistance to be obtained from the former drug has been formed by such writers as Freund, and although Elwin and Erlenmeyer have warned us not to fly from Scylla to Charybdis, still it is to be feared that the notion lingers that cocaine may be used advantageously and safely for this purpose. Nothing can be more mistaken. Cocaine is more seductive than morphia. It fastens upon its victim more rapidly, and its hold is, at least, as tight. Cocaine solutions are probably somewhat too freely prescribed in cases of diseases of the nose and naso-pharynx. Patients who use the drug in this way become very soon acquainted with its agreeable effects. Several cases have been recorded by American authors of cocaine habit arising thus. That cocaine has not been even more extensively misused is probably due to its being still a comparatively new drug, and also in part to its costliness. Up to the present time the largest number of its victims appear, unfortunately, to have been medical men.

Cocaine owes its special dangers to three causes. First, it is particularly treacherous. Secondly, it produces early mental breakdown both in the moral and intellectual spheres. Thirdly, it is intensely toxic, bringing about destructive tis-

sue changes after a comparatively short period of abuse.
Taking the last first, we know that alcoholic poisoning is
usually a slow process, while morphia may be taken even in
very large quantities for years without producing any serious
structural changes in the nerves.   In fact, we recognize no
distinct pathological results of morphia poisoning.   On the
other hand, the marasmus of chronic cocaine poisoning,
appearing early and developing with extreme rapidity, is
but one indication of the serious organic changes that are
produced.   Convulsions similar, as Richet points out, to
those of cortical epilepsy, have been noted in a great number
of cases.   In at least one recorded case, death occurred in
an epileptiform attack.   In animals, poisoned with cocaine,
remarkable rise of temperature has been observed by Mosso
Reichert and others.   Acute poisoning in animals kills by
asphyxia ; chronic poisoning, as Zanchevski shows, is accompanied by albuminous degeneration of the ganglionic cells
in the medulla oblongata and spinal cord, as well as of the
nerve cells of the heart ganglia and of the liver cells.   In
other more advanced cases this author has found atrophic
changes with vacuolation in the cells of the medulla and
cord, fatty degeneration of the muscular fibres of the heart,
and atrophy of the liver cells.   Degenerative changes also
occurred in the arterial coats, particularly in the spinal cord.
Perhaps organic changes similar, but less in degree, account
for the slowness and difficulty in recovering from the cocaine
habit, and the liability to dangerous collapse which exists
during the process of withdrawing the drug.

The treacherous and insidious character of cocaine
results from the fact that when taken in small doses it produces at first apparently nothing but a slight degree of exaltation, a sense of well being, a feeling of mutual and bodily
activity, of general satisfaction and of good humor that is
most agreeable.   There is no mental confusion which the
consumer of cocaine is conscious of, and the only overt
symptoms he betrays at this stage is more than natural

talkativeness. The hypnotic effects, when they appear, are not overwhelming, and there is no headache, no nausea, no confusion next day. This cocaine is probably the most agreeable of all narcotics, therefore the most dangerous and alluring. It is to be feared that these peculiar qualities may, indeed, conduce to raise this drug in the future to the bad eminence of being as Erlenmeyer says, the third great scourge of the human race (alcohol and opium being the first and second). Like several other observers, I have satisfied myself by experiments on healthy persons, that the agreeable results described actually follow the ingestion of small doses of cocaine, and this fact impresses one strongly with the feeling of how seductive this drug would be to the neurotic or debilitated. Of course, as is the case with all narcotics, small doses soon lose their effect, and hence a rapid increase is necessary.

The rapidity with which mental symptoms of a grave character appear is remarkable in cases in which increasing quantities of cocaine are taken. Within three months marked indications of degeneration, loss of memory, hallucinations,and suspicion deepening into persecutory delusions have been found.

The following is a characteristic clinical picture : on the one side in the cachexia or bodily ruin, on the other side in the moral impairment and pronounced mental affection. Patients who use cocaine alone—and those who have endeavored to wean themselves from morphine by its aid, and so added cocainism to the morphine habit—appear marasmatic. The skin is of a pale yellowish, almost cadaveric tint and withered feel ; the extremities are cool and covered with cold sweat. The eyes are deeply sunken, glistening, and surrounded by a dark ring ; the pupils widely dilated. Appetite is lost ; digestion disturbed. Salivation with dryness of throat may be complained of, and further, partial sensory disturbances or total analgesia. From the paralyzing action of cocaine upon the blood-vessels, patients complain of palpitation and breathlessness, troublesome sweating and noises in the ears, and also syncopal attacks and dyspnœa. The pulse is more frequent and easily compressible. They suffer from a want which must be satisfied ; they become nervous, trembling, and fall into a wretched condition of neurasthenia.

Speech is disconnected and can scarcely be understood ; impotence and incontinence of urine may appear. Sleeplessness sets in early. One of the most characteristic effects of this habit is the occurrence of muscular twitching, tonic and clonic convulsions, and finally epileptic attacks in which the patient may die. The mental symptoms may take the form of hallucinations, usually of general sensation, but not infrequently of sight as well. General mental weakness may set in rather early, to be observed in a loss of memory and unusual prolixity in conversation and correspondence. When the drug is withdrawn, besides the vasomotor symptoms there may be seen depression, impairment of will-power, weeping, etc. The chronic form does not protect from acute intoxication.

The treatment of a cocaine inebriate is practically the same as that of opium and alcohol cases. There is a degree of brain-degeneration apparent in the morbid impulses and strange, uncertain mental action of these cases that approaches very near to insanity. Abrupt withdrawal of the drug is the safest plan, and reliance is to be placed on bromides, foods, and baths to relieve the irritation and depression which follow. In the method of gradual withdrawal it has been found that small doses may result in collapse. In one case, in which 10 grains a day had been previously taken, and by gradual withdrawal the amount had reached 3 grains, symptoms of heart failure and paralysis suddenly came on. The dose was increased and the patient recovered. In another case, complicated with alcoholism, the daily allowance of 7 grains was reduced to 4, when a violent collapse set in, which was checked by giving the usual dose. The case finally became an alcoholic, but eventually recovered. The treatment of these mixed cases requires great care and watching, as they are especially liable to fatal collapse or to develop some form of acute brain disease.

Strychnine and the mineral acids are useful. Because of its rapid stimulant action strychnine, given hypodermically, 1-40 grain every three hours, will often prevent collapse when the cocaine is withdrawn. Mineral acids, chiefly phosphoric, may be given in 1 drachm doses three times a day as a general tonic, unless contraindicated by an acid stomach. In these cases mental conditions approaching delirium are present. The functional disturbance of the heart and the emotional changes that are so prominent in the latest stages, and the uncertain co-ordination of both the higher and lower brain and nerve-centres, all indicate the most serious disturbance. Some of the forms of iron, quinine, and strychnine may be very useful at the beginning of the treatment. Water containing bromide of sodium, 40 grains to the ounce, and charged with gas, can be given freely for some time with good results.

The therapeutics must be governed by the conditions of the case. All cases should be treated in an asylum, where the physician can have the full control of the conditions and surroundings of the patient. Home treatment is difficult and unsatisfactory. The patient is always hypersensitive to pain and suffering, and will use deception and intrigue to avoid it. The will-power is of no assistance, and the success of treatment will depend on accurate study and observation of the case. Drugs are of little value ; foods and hygienic care, including baths, will give the best results.

In all chronic drug-maniacs the physician should be on the lookout for cocainism. When cocainism is present a sudden addiction to opium, alcohol, chloral, or chloroform may arise, and in all cases the physician should be prepared for a sudden fatal termination of the case at any time.

CHAPTER XXXV.

CHLOROFORM INEBRIETY.

The following history of a case is so graphic and full of suggestions as to give the reader a good idea of such cases.

He says : " With me the chloroform infatuation was a case of love at first sight. I had been always temperate, almost a total abstainer from stimulants of all kinds. Once or twice I had smelled chloroform, and thought its odor pleasant. I was a young man just finishing my education, and fond of study. I had had some curiosity to know what it was like to be put to sleep with chloroform, and one night I happened to see a one-ounce bottle of chloroform which was bought for the toothache. I took the bottle home with me, and when I went to bed put a little of the chloroform on a handkerchief, and for the first time felt the delightful sensation of being wafted through an enchanted land into Nirvana. Those who know nothing of intoxication, except in the vulgar form produced by whiskey, have yet to learn what power there can be in a poison to create in a moment an Elysium of delight. It is a heaven of chaste pleasures. What I most remember is the vivid pictures that would seem to pass before my eyes—creations of marvelous beauty —every image distinct in outline, perfect in symmetry, and brilliant in coloring. The enjoyment is purely passive ; you have only to watch vision after vision ; but why each vision seems more wonderful and charming than the last you cannot tell, and you do not stop to question.

" I suppose that it was an unfortunate circumstance for me that I had never been drunk before in my life, and I never thought of comparing my blissful condition with that of the wretches I had sometimes seen staggering through the streets. I had made a great discovery. I had found a golden gate into dreamland—dangerous indeed to approach, I knew that, but who would heed any danger where the prize to be obtained was so great?—and guarding jealously my secret, I took care night after night to have by me the key to that golden gate. Probably I inhaled from half a drachm to a drachm or two each time. Generally I did not waken again until morning, and my sleep seemed to be just as refreshing as usual, only now and then I would wake with a trifling headache and feel disposed to lie a little longer in bed than common. My bodily condition did not seem to suffer in the least, and my faculties all seemed as keen as ever. I felt no craving for my pet intoxicant during the day—did not give it a thought often until bed-time came, and then it would occur to me for a moment to try and see how it would seem to go to sleep in the ordinary way, the conclusion always being that— to-morrow night I would make the experiment. So, before I knew it, I was a slave. I would say to myself, 'It does not hurt me ; it seems to have no more effect than the cigar my friend smokes after dinner. Really I believe it is a positive benefit. It seems to keep my bowels regular, and it certainly makes me sleep soundly all night.'

"But after a while I found that I was using a larger quantity of chloroform than at first. I would take a two ounce bottle half-full of the stuff to bed with me, and inhaling directly from the bottle would forget at last to cork it, and in the morning it would be empty. Sometimes I would wake after midnight, or partially wake, to take another dose. I found that there was a bad taste in my mouth all the time, keeping me in mind of chloroform. I was often nauseated in the morning, and sometimes at intervals during the day. I began to feel a longing for chloroform whenever I had a little headache, or was dispirited from any cause, and I sometimes yielded to what I already knew was a morbid craving. I began to be indifferent to the things that personally had interested me, avoided society, and became depressed in spirits. My complexion became sallow, whites of the eyes yellow, the bowels sometimes windy and unnaturally loose, skin dry and seemingly bloodless, and injuries of the skin did not heal rapidly. In winter there was a tendency to chapping, that had not before been noticed.

"Meanwhile I had ceased to have visions, or they came rarely. I began to realize that my pet habit was becoming my tyrannical master. I had no special cares to drown, but it became my insane pleasure to draw over my senses the veil of oblivion. I loved the valley of the shadow of death. I knew there was danger that some night I should pass over the line, into a sleep from which there would be no waking ; but death had no terrors for me. Nay, to bring all my faculties and powers and ambitions into the sweet oblivion of transient death was the one pleasure for which I cared to live. I was conscious of a profound moral deterioration ; I became materialist ; I had no soul ; immortality was a dream of the ignorant ; I, who had a thousand times annihilated my own soul with my senses, knew that the dream had no corresponding reality.

"Yet all this time I continued faithful in my daily duties, and resisted successfully the temptation to hurry through my evening so as to get the sooner to my chloroform. I did not admit to myself that I was a slave to the habit, or even that the habit was an injury to me, as yet; but I began to be afraid, and the more when I found, when I resolved (as often I did), to omit my nightly indulgence, just for a week, how impotent my will was in the matter.

"This was my condition at the end of two years. I was still only using a moderate quantity of chloroform, about three drachms daily, exceeding that quantity only by accident. An opportunity offered for a change of occupation and surroundings, which I eagerly seized in the hope that it might enable me to break my fetters. For about three months, under the new surroundings, I abstained from chloroform, and found it really not difficult to do so. I began to think that I had greatly over-rated the power of the habit. At all events, after the first week I had no craving for the stimulant. But one day I came across a bottle of chloroform. When I saw it I smiled to myself to think that I had imagined myself a slave of any such thing. Night came, and when I was ready for bed the devil of appetite gave me his commands, and I obeyed. Just one smell to see whether I really wanted it ; I would not take the bottle to bed with me. So I inhaled, standing, directly from the bottle—a full pound of chloroform—and with the first breath of the vapor came back, with renewed force, all the old appetite, keener than ever from long abstinence. Once more I saw the old-time visions, as beautiful and as vivid as at first. One peculiarity of these visions I may speak of right here. Objects would appear with wonderful sharpness of outline just as they would be seen with the eyes, only reduced to micoscopic size like objects seen through an inverted microscope.

"To go on with my story. What happened after I got the bottle in my hands I do not know. The next morning I found the bottle corked and in its place, but only half full of chloroform, and I was told that I had been lying in some kind of a fit ; some thought I was drunk—as indeed I was. From this time I realized myself a slave, but not now a willing one. I did not again commence at once the use of the chloroform, but at intervals of from three to eight weeks would indulge in a regular spree, lasting from one to three days, during which I would keep myself as nearly as possible dead drunk, and would consume from four to eight ounces of chloroform. All this time I kept my habit a secret, and continued to do my ordinary work with the usual zest in the intervals between my sprees. At last discovery came. You well remember how I was found apparently lifeless, and how by the active use of restoratives you brought me to myself. How my moral perceptions were quickened the moment I saw myself through the eyes of another!

"You know that it was not in a week or a year that I was placed morally on a firm foothold again. Indeed, you did not know how often, after I had given you and myself my word and pledge to abstain wholly from chloroform, I relapsed, taken unawares by the tempter. For more than two years I kept up the conflict, too often thinking the final victory won, only to find there was one imperative command it was useless for me to attempt to disobey, and that command came to me whenever the least whiff of chloroform entered my nostrils. Once or twice I tried the expedient of returning to my first practice of a regular moderate use of the stimulant, but I found that moderation was now almost impossible. If I went to sleep under the influence I would awake again, and find myself then unable to sleep, distressingly wide awake and nervous, until I courted again my 'dearest foe.' Symptoms like those of delirium tremens several times developed. I saw ' things,' not now beautiful visions, but shadowy images, that filled me with nameless, irrational horror. Appetite was capricious. I was frequently nauseated, but food seemed to relieve this condition ; vitality was low, the blood ran sluggishly in my veins, and seemed especially to desert the surface of the body. I suffered particularly in cold weather, and it was during cold weather in winter, especially, that I found it almost impossible to resist my besetting temptation.

"At last I prevailed by sheer force of will. I had recovered enough faith in the soul to assert my freedom, and I now look back upon those years of conflict with a kind of self-pity, to think I could have been so weak. But I do not to-day court temptation. I am not conscious of a lurking appetite, but I dare not put my virtue to any severe test. I am sure, however, that the chloroform habit is one that can be broken by steady determination. I have no faith in any process of tapering off. It is just as easy to quit once for all as to prolong the agony, and the suffering is often purely imaginary. It took many months for me to recover. If doctors only knew the fascination of this drug, they would seldom or never prescribe it. The danger of the wine cup is nothing to that of the chloroform bottle."

CHAPTER XXXVI.

COFFEE AND TEA INEBRIETY AND THEIR EFFECTS.

Most physicians are doubtless able to recall numerous instances in which coffee has induced more or less serious symptoms. It seems that personal idiosyncrasies often determine the extent of the evil. The evils upon the eyes and ears of people are more frequent from coffee than from tobacco or alcohol. It does not absolutely destroy vision or hearing, but it induces functional troubles very annoying to their possessors. That coffee is the efficient agent appears from the fact that upon the entire discontinuance of the use of coffee, the symptoms complained of disappear.

Dr. Guelliot has published twenty-three cases of chronic caffeism. Of these cases seventeen were women.

The following are the principal symptoms :

Anorexia, disturbance of sleep, trembling of the lips and tongue, attacks of gastralgia, different kinds of neuralgia, dyspepsia, and leucorrhœa, often profuse. In the twenty-three cases, he found in eighteen, anorexia ; in sixteen, disturbance of sleep ; in sixteen, trembling of the lips and tongue ; in twelve, leucorrhœa ; in eleven, gastralgia ; in ten, dyspepsia ; in ten, neuralgia of various forms ; in eight, cephalalgia ; in four, vertigo and convulsive attacks ; in four, obstinate constipation ; and in three, constipation and diarrhœa alternating.

The patients had pinched, pale, wrinkled faces, a weak, rapid pulse, and the sleep was disturbed by anxious dreams. The following is the account of a typical case : A woman

in middle life kept her pocket full of coffee, which she ate
constantly.   Her skin was of an earthy tint, constipation
was obstinate, sleep very irregular, and her mind restless,
anxious, and full of forebodings.   She was much emaciated,
and both the nervous system and digestion suffered severely
at times.   The lips and tongue were tremulous, dry, red,
and cracked.   The appetite was very irregular, and vertigo,
prolonged headache, and epigastric pain were present most
of the time.   She was placed under treatment and became
delirious.   Beef-tea, milk, baths, and a mild galvanic current
were used for several weeks, and these were followed by
bark tonics.   She was discharged restored six months
later.

The evil effects of coffee are especially observable in
children.   The coffee drunkard is described as thin, pinched
features, pale, wrinkled face, and a grayish yellow complexion.   The pulse is weak, frequent, and compressible.   The
sleep is troubled with anxious dreams.

Although coffee does on the whole far more good than
evil, it is important to bear in mind the evils that it is able
to produce under favoring circumstances.   In a general way
it may be said that indoor brain workers do not bear coffee
as well as outdoor muscle workers.   Persons of nervous
temperament bear coffee badly.

The effects of coffee when pushed to an excess may be
to some extent confused by the alcohol and tobacco which
often accompany it, but they can be studied more accurately
in women, especially in those who do not drink coffee, but
eat it.

As a rule both nervous system and digestion suffer in
these cases.   The appetite fails, there are attacks of sharp
epigastric pain, much vertigo, and prolonged headache.
There is less insomnia than might be supposed by those
who know the weakening power of a single cup, but much
dreaming and restlessness of a non-aphrodisiæ type.   The
pulse is weak and quick, there is often an anæmic mur-

mur. The muscles waste quickly. The coffee inebriate is always thin. He may be a mere skeleton ; his eyes are bright and quick in movement, their pupils large ; and may be mistaken for a tea-drinker. In the insomnia which follows, when the coffee is removed, the only remedy is the old poison.

As with tea-drinking, coffee addiction is followed by the employment of spirits and other drugs. Many inebriates and opium-takers have a history of excessive use of coffee before the other drugs were taken. The recognition of addiction to coffee is important in many cases of neurotics, especially in children and young persons, and unless promptly checked will be followed by serious results. The excessive use of coffee in all cases is a very significant hint of nerve-exhaustion and disorder of the motor nerves.

In the late war many cases of delirium from coffee were noted where the food-supply was scant and coffee was abundant. Some of these cases came under special treatment, and yielded readily to baths, mineral waters, and strong foods. When coffee seems first to have been used for insomnia, the treatment must depend on a careful study of the etiology, and from the removal of the causes the cure may be expected. I have also noted a number of cases in young children of inebriate and neurotic parents who developed a morbid impulse for coffee. Such cases require active treatment, and milk, mineral waters, and baths are prominent remedies. Neurotic disturbances and diseases from coffee are but little known.

Dr. Mendel, of Berlin, has lately published a clinical study of this neurosis, which is growing rapidly in this country. His observations were confined to the women of the working population in and about Essen. He found large numbers of women consumed over a pound a week, and some men drank considerable more, besides beer and wine. The leading symptoms were profound depression of spirits, and frequent headaches, with insomnia. A strong dose of coffee would relieve this, for a time, then it would return. The muscles would become weak and trembling, and the hands would tremble when at rest. An increasing aversion to labor and any steady work was noticeable. The heart's action was rapid, irregular, and palpitations and a heavy feeling in the precordical region were present. Dyspepsia of an extreme nervous type was also present. Acute rosacea was common in these cases. These symptoms constantly grow worse, and are only relieved by the large quantities of coffee, generally of the infusion. In some cases the tincture was used. The victims suffer so seriously that they dare not abandon it for fear of death.

Where brandy is taken only temporary relief follows. The face becomes sallow, and the hands and feet cold, and an expression of dread and agony settles over the countenance, only relieved by using strong doses of coffee. In all these cases, acute inflammations are likely to appear any time. An injury of any part of the body is the starting point for inflammations of an erysipelatous character. Melancholy and hysteria are present in all cases. In this country the coffee-drinker after a time turns to alcohol and becomes a constant drinker. In other cases opium is taken as a substitute. Coffee inebriates are more common among the neurasthenics, and are more concealed, because the effects of excessive doses of coffee are obscure and largely unknown. Many opium and alcoholic cases have an early history of excessive use of coffee, and are always more degenerate and difficult to treat. A very wide field for future study opens up in this direction.

Dr. Slayter describes a case of delirium in a girl who chewed large quantities of tea. It appeared that masses of tea leaves had lodged in the bowels, and the delirium was in some measure dependent on the irritation and reflex action which followed. Trembling delirium, and delusions of injury from others, gave it a strong resemblance to delirium tremens. The amount of tea chewed daily was over one pound. The patient recovered by the use of free cathartics and the withdrawal of the tea. In 1881 I saw a boy who had delirium and trembling that had existed at intervals for two months. The fact that his father had died an inebriate seemed to be a sufficient reason for his symptoms in the minds of his friends. It was ascertained that he had for years drank large quantities of tea. Having been employed in a tea-store, he had chewed it freely. He was literally a tea inebriate. He had inherited an inebriate diathesis, and the early and excessive use of tea was a symptom of it. He had all the symptoms of one who was using alcohol to excess. He recovered, and a year later used coffee to great excess, until he became unfit for work ; then was under medical care for a time, recovered, and finally became an opium-taker.

Another case came under my observation in the person of a little girl, twelve years old, the daughter of a patient under my care for inebriety. She had gradually and steadily become excessively nervous. Could not sleep, had muscular twitchings and delusions of fear ; would burst into tears, and complain that she was going to be turned out into the streets. She heard voices at night, and could not keep still. She also imagined that her father was being burned. It was finally found that she was a tea inebriate, and both drank and chewed it at all times and without any restraint. A physician consulted me about a singular stage of trembling and mild delusions which had appeared in a family of three old maids living alone in the country. It was found to come from excessive use of tea, and to be tea inebriety. When this was stopped they recovered. My observation leads me to think that these cases are not uncommon among the neurotics. They are of such a mild character at first as to escape special observation, and hence are supposed to be due to other causes. Such cases, after beginning on tea, take other drugs, and become alcohol, opium, or chloral takers, or develop some form of neurosis, which covers the real and first causes.

Theine is the active principle of the leaves of Chinese tea, and is generally reputed to be identical with caffeine, both in chemical composition and in physiological action. My experiments show that it differs very markedly in physiological action from that of caffeine. Caffeine principally affects the motor nerves, while theine chiefly influences the sensory nērves, and clinically proves itself a most valuable analgesic, surpassing morphia in promptness and permanency in relieving pain in some affections, without producing any, or at least very little, disturbance of the general nervous system. It paralyzes sensation before motion; it impairs sensibility from the centre to the periphery and not, like brucine and cocaine, from the periphery to the centre; it produces convulsions which are spinal and not cerebral; it has a more powerful action on the sensory nerves, and less on the motor nerves than caffeine.

From the results of theine in these cases it will be seen that it is a powerful anodyne without producing any intoxication of the higher nerve centres, which is so common with morphia and all other agents belonging to this class. Its influence is both quick and persistent, and it manifests an almost exclusive affinity for the sensory nerves. It relieves pain by acting from the centre toward the periphery and showing its effects but very seldom above the seat of injection. In 1-10, 1-5, and even ⅓ grain doses it is entirely free from dangerous consequences—the only inconvenience which it causes is a slight, but transient burning at the point of introduction. I use a one per cent. watery solution of Merck's preparation—ten minims of which equal one-fifth of a grain of theine. Larger doses are required in some individuals in order to bring out its characteristic action.

EFFECTS OF TEA DRINKING ON THE NUTRITION OF THE EYEBALLS.

Dr. Wolfe has described the first effect as one of softening of the vitreous humor, which became filled with floating particles of pigment. It had come under his notice in persons who at first sight seemed to have very little in common. He had found it among—1. The mining population, who pass a deal of time underground. 2. Washerwomen. 3. Middle-aged laborers, masons, and out-door workers. 4. Shop and factory girls. 5. Not a few belonging to the upper classes. His attention was specially directed to the affection by its frequent occurrence among Australians who came to consult him. He could discover no assignable cause for the disease, either in the tissues themselves or in the history of the patient ; and it was only on directing his inquiries to their diet, and finding that they all agreed in consuming large quantities of tea, that he came to suspect its agency. A comparison of the numerous cases of opacity of the vitreous humor occurring among tea-drinking populations, with its less frequency in France, Germany, and America, and its rarity among the Turks, tended to confirm his suspicions. Physiology did not suggest an explanation, but chemistry pointed to theine and tannic acid as most likely to cause disease. Theine might be left out of consideration, being identical with caffeine, which was innocuous ; so there only remained tannic acid. This precipitated albuminoids from their solutions ; hence it probably acted injuriously by precipitating some . of the most important constituents of the food, and also by affecting the mucous membrane of the stomach and alimentary canal, and thus preventing digestion and assimilation.

Some observations had been made as to the effects of tea-drinking on the healing of wounds and ulcers, by a Glasgow surgeon, who had noticed that, in persons addicted to this habit, they took on a sort of scorbutic character. Physicians also ascribed numerous cases of rebellious dyspepsia to the use of tea. The disease of the vitreous humor above alluded to, could hardly be an isolated pathological fact, but must be associated with deleterious changes in other parts of the economy, and probably only made its appearance in organs which had a predisposition to be so affected. Without venturing upon any theory as to the action of tea on the vitreous humor, he would point out that the first expression of acute irritation of the fifth nerve in sympathetic ophthalmia was opacity of the vitreous humor and detachment of pigment from the whole uvular tract. So it was possible that chronic irritation of the same nerve might give rise to such changes in the nutrition of the eyeball as to bring about the condition under consideration. He commended this subject to the notice of general practitioners, who had better opportunities of judging of it than he had.

CHAPTER XXXVII.

ON PSYCHOSIS CAUSED BY NICOTINE.·

Nicotine is the most important chemical substance contained in the West Indian plant, *nicotiana tabacum ;* when pure, it is a colorless, easily soluble fluid, of strong tobacco odor and very acrid, burning taste, is easily dissolved in water, alcohol, and ether, its reaction being strongly alkaline, and it forms simple crystallizing salts. It is said that Virginia tobacco contains the largest quantity of nicotine. Nicotine acts on the human organism as one of the most powerful poisons ; the action of tobacco differs only in degree.

From experiments instituted in Professor Schroff's laboratory it results that the effects of nicotine on the healthy organism are as follows : Taken in doses of 1 to 3 milligrams, the alkaloid produced first an acrid burning sensation on the tongue and in the throat, with increased salivation, and a sensation of heat in the stomach, in the extremities, and in the whole body. Soon after this the phenomena are headache, vertigo, drowsiness, impaired vision and audition, accelerated and oppressed respiration. They were succeeded within half or three-quarters of an hour after ingestion of the poison by an extraordinary feeling of relaxation and weakness, the face became pallid and the whole body as cold as ice ; fainting fits and vomiting made their appearance. The symptoms increased to tumultuous and chronic spasms of the respiratory muscles. After three hours the symptoms commenced decreasing, but secondary effects continued for several days.

Subsequent experiments demonstrated that the first action of the poison is excitation, the ulterior action is depression and paralyzation. In the beginning the functions of the brain and of the spinal cord are enhanced, but this excitation is followed by relaxation and debility. The whole voluntary muscular system is subjected to this influence, which gradually extends to the heart, and finally to the vasomotor system. Such are the effects of this redoubtable poison on the healthy organism, when taken in small doses of 1 to 3 milligrams. What may be the effects of the 1,200,000 kilograms of the same alkaloid, constituting the present amount produced on the whole surface of the globe? It is impossible to ascertain in an exact manner the number of men using tobacco, but it is asserted that approximately they number 800,000,000. Each of them consumes on an average 1 ½ grams nicotine every year, or 4 milligrams every day, some of them less, others considerably more. The fact that large doses of this amount are supported by man proves once more the adaptability of the human nervous system to injurious habits.

The first contact with the tobacco poison is always felt by the cerebral nervous system as painful and hostile, and its repetition requires a certain violence against nature. By not heeding this warning of the attacked nervous cell, and by repeating the essays with the necessary perseverance, the repulsive sensation gradually vanishes and the excitation remains as a stimulus, which soon becomes indispensable. At last the irritation and debilitation of the cerebral nervous system reaches a degree in which privation is deemed a real suffering, and the same longing for renewed enjoyment of the poison manifests itself as is noted in alcoholists and morphinists.

The action of nicotine differs considerably, not only according to the different classes of tobacco, but also according to its different applications. Smoking only allows the products of decomposition connected with it to exercise an influence, although several cases are known, not only of nervous erethism, but even of perfect psychosis, caused by excessive tobacco smoking.   Kjellberg had several opportunities of convincing himself that even a too abundant use of tobacco for snuffing may originate psychic phenomena.

But the greatest danger lies in the use of tobacco for chewing.   The ordinary shape for this application is tobacco in rolls.   It is true that small doses of it may be taken without causing psychic injuries.   But as soon as the daily use exceeds 10 to 12 grams of genuine and good tobacco, the field of pathologic phenomena is approached, and the sensation becomes an abnormal one.   This transition is effected more rapidly in case pulverized snuff is used for chewing.   In the last decades the use of snuff for chewing has considerably increased in various northern countries, especially among mariners, manufacturing laborers, etc.

From a number of cases observed by him, and in which a continued daily use of from 20–27 grams of tobacco in rolls, or in snuff, had been ascertained, Kjellberg has reached the conclusion that the similarity of symptoms points to a specific psychosis caused by the use of tobacco, " *nicotinosis mentalis*," a real primary mental disease with its own peculiar symptoms, which are clinically distinguishable, and which give it a place among mental intoxications. " *Nicotinosis mentalis* " is described by Kjellberg as follows :

Among general symptoms a painful sensation of weakness and impotence is to be noted, accompanied very soon by hallucinations, maniacal ideas and suicidal inclination. The disease has a prodromic stage and three distinct stages differing from each other.

Prodromic stage—The patient has felt unwell for some time, his general disposition changes, he shows an unusual uneasiness which may pass to a transient state of anguish. He sleeps little, and the ordinary occupations are repugnant to him.   He is disposed to indulge in somber reflections, is tormented by palpitations of the heart and unusual anxiety.

This condition continues for one and one-half and three months, when psychosis sets in.

First stage.—The patient's attention is seized by lively hallucinations, and he is entirely occupied with these new and surprising perceptions. He hears voices, visions appear to him, he has a sensation as though something different from himself was inside his body, and strange notions take hold of his conscience, from which he can't free himself. False, fixed ideas often combine suicidal tendencies; his mood is always gloomy; the patient feels tired and exhausted, inclined to loneliness and rest; at times he has short fits of fright. Otherwise he is quiet and obedient, talks little, and never without being previously requested to do so, but then what he says is logical and shows good perception. He complains of painful sensation in the heart, of wearisome insomnia and of voices that give him no peace. Nutrition is not impoverished, for the patient usually eats plenty, although he often speaks of bad appetite and spoiled food. After six or seven months the disease enters a new stage.

Second stage.—The mental disposition improves, and we find the patient talks hilariously of his perceptions. He relates the visit of angels, he has seen heaven, but also hell and the evil spirits. He sings and talks to himself without interruption in a low voice; motions become more vigorous, and he moves about with a certain agitation. The hallucinations of brain and vision are on the increase, and the patient is constrained to execute involuntary motions. This condition is periodic, the periods usually lasting from two to four weeks with intervals of indefinite length. In such times the patient lays down prostrated and lazy, his mood is gloomy and discontented, his attention diminished, his perception very slow, his language distinct and logical, but hesitating. This stage may continue for a long time, but unless convalescence sets in, it passes gradually into a last stage.

Third stage.—The intervals pass each other and periods of raised disposition disappear by degrees; the mental disposition remains quiet, perception is very limited. Hallucinations continue, and the patient gradually sinks into a state of general psychic debility, while his physical condition improves, and he may even be partially serviceable for ordinary occupations.

With regard to prognosis, it is not entirely bad during the first and second stages. But in the third stage recovery is not to be expected any more. Therapeutic treatment requires, first of all, absolute deprivation of tobacco, with the understanding that it should be enforced by degrees, the patient being otherwise subjected to very painful sensations and much suffering. With this, substantial diet, motion in the open air, and use of mineral waters are to be recommended.

CHAPTER XXXVIII.

EAU DE COLOGNE DRINKING AND ARSENIC TAKING.

The use of cologne as a substitute for spirits is very common among inebriates ; generally when no other form of spirits can be procured. Recently attention has been turned to the rapidly increased consumption of cologne, both in large cities of Europe and this country, and the conclusion reached by several authorities is that cologne is becoming a drink in many circles in preference to other forms of spirits. To many persons this odoriferous compound is very attractive, and especially when the cologne is made with methylated spirits its spirit strength is equal and exceeds many of the stronger alcoholic drinks in market.

The factories for its manufacture in Cologne use the following general recipe : Twelve drops of the essential oils neroli, citron, bergamot, orange and rosemary ; one drachm of malabar cardamons to one gallon of rectified spirits. In this country cheap wood spirits are used, which gives greater alcohol strength at half the expense.

In England many women and men in the better walks of life begin by taking a few drops of this perfume on sugar in the morning for some debility. This increases until they come to depend upon it the same as any other spirit compound. It can always be purchased with ease and without exciting suspicion, and can be used with great secrecy. American cologne is most often made from wood spirits, and is a very fiery, nauseous mixture, which cannot be taken in water, but used on loaf sugar or lozenges is endurable.

383

In one instance a wine glass of cologne was taken three or
four times a day for a long time ; the patient finally died
from delirium tremens.   It appears that the effects of this
drink vary but little from ordinary strong spirits, except,
perhaps, there may be more profound nutrient disturbances,
insomnia, and tendency to delirium. If the cologne is made
from wood spirits, the brain and nerve degeneration is both
intense and profound, and delirium is very sure to follow.

It has been asserted that melancholia and insomnia in a
case suspected of using spirits in secret is an indication of
the use of cologne. Usually the cologne-drinker will have
a strong odor of this perfume about his body and breath
which cannot be mistaken.   Such cases usually use this perfume externally in excess to divert suspicion from its
internal use.   Undoubtedly there are, in this country, an
increasing number of cases where cologne is used secretly
and exclusively. These cases, no doubt, become morphia,
chloral, and cocaine inebriates after a time, and in some
instances from a physician's prescription which contains
these drugs, that are often fascinating substitutes.   The
alcohol and opium inebriates turn readily to cologne, and
use it freely and with great satisfaction.

The American inebriate, if a man, is not likely to use
this perfume very long as a drink, but if a woman, it may
be taken for years in secret. Obscure and complex
nervous disorders in a woman that uses cologne externally
should always suggest the possibility of its internal use.
Inebriates who use it externally and recover rapidly, or
make sudden changes of habits and living may be suspected of substituting it for other spirits. Cologne, both
German and American brands, contains a large and vari-
able per cent. of alcohol, and are always dangerous for use
among neurotics, even externally.   Its internal use is very
likely to follow if the person has a great liking for this perfume. In hospitals for the treatment of alcohol and opium
cases cologne is found to be as dangerous as alcohol and is

not allowed. In private practice among neurotics the possibility of this danger should always be considered. It is
asserted that the sale of cologne has increased enormously
in certain sections of this country. Statistics on this point
would be very interesting. We trust the coming census
will throw some light on this, and the extent of the use of
bitters in this country.

ARSENIC INEBRIATES.

In the highlands of Styria the prevalence of inebriety is
limited, if not directly counteracted, by the strange habit
of arsenic eating. Beginning with a minimum quantity,
the devotees of the baneful drug gradually accustom their
organism to a dose that would prove promptly fatal to any
non-habitué, and persist in excusing their practice by all
sorts of sophisms. Arsenic counteracts the raw air of the
bleak uplands, it enables travelers to resist the fatigues of
mountain-climbing, it stimulates digestion, etc., etc.

"South of this city" (Gratz), writes an Austrian physician, "arsenic eaters are found in nearly every village;
there are families where the drug is used by every male
adult, and often by white-haired patriarchs of seventy or
eighty years, for it must be conceded that the habit is not
incompatible with longevity. As a rule, I have been able to
recognize a poison-eater by a certain moody appearance,
contrasting strangely with the jovial disposition of the high-
landers in general. A sallow, though clear and wax-like,
complexion, is another characteristic symptom, and, in large
doses, the use of the drug often involves serious digestive
disorders.

"It must be admitted, however, that arsenic never produces anything like intoxication in the uglier sense of the
word. The poison-eater directly after a large dose, feels
elated; his habitual moodiness gives way to a more
buoyant disposition, but he does not become quarrelsome

or maudlin-sentimental ; he can keep his temper in a lively controversy and the clearness of his intellect is not visibly affected. Nor are his financial circumstances apt to be imperiled by the habit. At retail rates, half a florin (twenty-five cents) worth of white arsenic will keep a whole family in stimulants for a couple of months. Very few inveterate poison-eaters can use up a florin's worth in the course of a year, while brandy drinkers often waste the wages of a week's hard work in the mad revels of a single night. And while one such night incapacitates or indisposes the toper for work during the next forty-eight hours, an arsenic eater, under the full influence of his tonic, and for hours after, can follow his usual occupation as if nothing had happened."

INEBRIETY FROM GINGER DRINKING.

The increasing demand for ginger extracts and drinks is a very significant hint of a new phase of the morbid drink impulses of the age. Several large proprietary establishments are devoted exclusively to the preparation of ginger extracts, essences, and drinks, which are extensively advertised as medicines and preventive drinks for the diseases of the different seasons. It is a well-known fact that all these preparations are made with the poorest, cheapest spirits, and contain from thirty up to eighty per cent. of alcohol. In some instance wood spirits are used on account of the cheapness, and the intoxicating qualities of this mixture are far worse than any alcoholic drinks of commerce. The demand for these ginger drinks is due to the alcohol they contain, the ginger in itself having but little influence on the body, although some enthusiastic writers assert that ginger taken in large quantities produces a distinct form of inebriety, marked by stupor and melancholy.

In two cases which have been reported where extract of ginger was taken in large quantities, profound nutrient disturbances, and inanition were present. The intoxication was less maniacal, and attended with profound depression. This would undoubtedly depend on the alcohol more than the ginger. From inquiries it appears that there are a large number of persons who buy extract of ginger regularly, apparently using it as a common drink. The probability is that after a few months or years they abandon this drink for some stronger alcoholic drinks, or narcotics.

A New York druggist writes : "That the sale of ginger extracts to women are rapidly increasing ; that he has over a dozen regular customers, who buy from two quarts to one gallon of ginger a week."

The sale of ginger in Maine was so great, that it was declared by the courts to be an intoxicant, and placed among the alcoholic drinks prohibited.

From a variety of evidence there can be no doubt that ginger drinking in this country has reached a dangerous magnitude, and those who use it any length of time are almost certain to become alcoholic or opium inebriates.

The extracts of ginger on the market are without exception dangerous, because of the dangerous alcohols they contain. Neuræsthenics and neurotics should avoid them as poison, and inebriates of every form will always find them treacherous remedies for every condition. For all the various functional disturbances they are supposed to relieve, pure alcohol is far safer, and less injurious.

The following item is of interest :

An officer of a church accused the pastor of being a ginger inebriate. A trial followed and Dr. Day was called as an expert. The peculiar conduct of the clergyman, with the great number of extra ginger bottles found in his private study, with other evidence, pointed to the conclusion that he was a secret ginger inebriate. Dr. Day said that for a long time he was convinced that the use of alcohol, plain and simple, was being superceded by a preference for new-fangled nerve tonics that possessed substantially the essen-

tial properties of alcohol, without the disagreeable reputation, and to discover what peculiar forms this new
manifestation was taking.

For thirty-one years he had been treating nearly one
hundred cases yearly that displayed all the symptoms of
alcoholic delirium when the cause of disease was not,
strictly speaking, alcoholic indulgence.

Besides chloral and cocaine inebriates, of which we
always have more or less, I had a man who used to get
furiously drunk on quinine. He'd stagger about and yell
like a regular whiskey drunk, and he'd go through all the
successive stages of a common intoxication. He'd get jovial
and pleasant first, and then cross and savage, and finally
maudlin. He was a very interesting case. I have seen
some cases suffering from the effects of the use of patent
medicines ; for instance. Many of these nerve-tonics and
so-called patent foods are cultivators of an intemperate
habit that results in mental and physical manifestations
very similar to the indications of excessive alcoholic indulgence. A well-known proprietary food was nothing better
than an alcoholic stimulant.

In regard to Jamaica ginger : "It takes the strongest
kind of alcohol to preserve Jamaica ginger, and the tincture of this substance is extremely inebriating when used
even in small quantities. I knew a patient who used to
get drunk on a spoonful or two of this stuff, and there isn't
any doubt that a great many people use it as an intoxicant.
Anything almost acting as a stimulant to the nerves may be
used as an intoxicating agent. I knew a man who got
quite drunk on a strong cigar, and I've had several cases
of physical derangement of pronounced alcoholic appearance from excessive smoking. Tobacco and alcohol seem
to have a very close identity. Many a drunkard will rest
comparatively content when deprived of his liquor, if a
good supply of tobacco is furnished him, and many an
inveterate tobacco user falls into drink when he is deprived

of the favorite weed. I knew a confirmed tea-drinker who was badly shattered from the excessive indulgence in what the poet calls "the cup that cheers, but not inebriates," and there are instances of mild intoxication resulting from too much tea-drinking.

"The tendency of these days seems to be toward new forms of inebriety, and the effect of alcoholic indulgence now is much more injurious, because the practice of adulteration has become so extensive. Then the odium which attaches to the use of alcohol has led many people in high places to cultivate a habit of secret tippling, and such people frequently devise an original method of securing the effect of inebriety without subjecting themselves to the charge of using alcohol. Consequently the use of outlandish and sometimes deadly drugs is clearly on the increase."

APPENDIX.

DRUNKENNESS AS A DEFENSE.

BY CLARK BELL, ESQ.

The law as now settled in England and the American States may be stated as follows :

While drunkenness is not *per se* a defense upon a charge of crime, yet mental unsoundness, superinduced by excessive intoxication and continuing after it has subsided, may excuse ; or where the mind is destroyed by a long-continued habit of drunkenness ; or where the long-continued drunkenness has caused an habitual madness, which existed when the offense was committed, the victim would not be responsible. For if the reason be perverted or destroyed by a fixed disease, although brought on by his own vices, the law holds him not accountable : Rex v. Meakin, 7 Car. & P., 297 ; Reume's Case, 1 Lewin, 76 ; Reniger v. Fogassa, Plow., 1 ; 1 Russ. on Crimes (9th ed.), 12 ; 1 Bishop Cr. L. (6th ed.), 406 ; 1 Wharton Cr. L. (8th ed.) sec. 48 ; McDonald C. L. of Scott., 16 ; 1 Hale, 4 ; Black. Com., 26 ; Beasley v. State, 50 Ala., 149 ; Peo. v. Odill, 1 Dak. Ter., 197 ; Estes v. State, 55 Ga., 30 ; Baily v. State, 26 Ind., 422 ; Roberts v. People, 10 Mich., 401 ; s. c., 19 Metc., 402 ; State v. Hundley, 46 Mo., 414 ; State v. Thompson, 12 Nev., 140 ;

Lanergan *v.* People, 50 Barb. (N. Y.), 266 ; Maconnehey *v.* State, 5 Ohio, 77 ; Com. *v.* Green, Ashm. (Pa.), 289 : U. S. *v.* Forbes, Crabbe (D. C.), 558 ; Stuart *v.* State, 57 Tenn., 178 ; Carter *v.* State, 12 Texas, 500 ; Bell's Med. Jurisp. of Inebriety, p. 10, and cases there cited).

The rule of law is well settled that evidence of intoxication is always admissible to explain the conduct and intent of the accused in cases of homicide, although the rule does not apply in lesser crimes, where the intent is not a necessary element to constitute a degree or phase of the crime. Bell's Med. Jur. of Inebriety, p. 10, and cases there cited.

In cases where the law recognizes different degrees of a given crime, and provides that willful and deliberate intention, malice, and premeditation must be actually proved to convict in the first degree, it is a proper subject of inquiry whether the accused was in a condition of mind to be capable of premeditation :   Gray. J., in Hopt *v.* People, 104 U. S., 631 ; Buswell on Insanity, § 450 ; Penn *v.* McFall, Addison, 255 ; Keenan *v.* Commonwealth, 44 Pa. St., 55 ; Jones *v.* Com., 75 Pa. St., 403 ; State *v.* Johnson, 40 Conn., 136 ; Pirttle *v.* The State, 9 Humph., 663 ; Haile *v.* State, 11 Humphrey, 154 ; Smith *v.* Duval (Ky.), 224 ; Bosswell *v.* Com., 20 Gratt., 860 ; Willis *v.* Com., 32 Gratt., 929 ; People *v.* Belencia, 21 Cal., 544 ; People *v.* King, 27 Cal., 507 ; People *v.* Lewis, 36 Cal., 531 ; People *v.* Williams, 43 Cal., 344 ; Farrell *v.* State, 43 Texas, 508 ; Colbath *v.* State, 2 Tex. App. 391 ; State *v.* White, 14 Kan., 538 ; Schlacken *v.* State, 9 Neb., 241 ; 104 U. S.

The reason of this rule of law rests upon the fact that intoxication is a circumstance to be weighed in connection with the other circumstances surrounding the commission of the act in determining whether it was inspired by deliberate and malicious intent, and whether immediately before and at the time of his act the intoxication of the accused was so great as to render him incapable of forming a design or intent, which the jury must find from the facts in the case, without regard to the opinions of others: Buswell on Insanity, § 452 ; Marshall's Case, 1 Lew. Cr. Cas., 76 ; Thacher, J., in Kelly *v.* State, 3 S. & M., 518 ; Armor *v.* State, 63 Ala., 173 ; People *v.* Belencia, 21 Cal., 54.

And because, since he who voluntarily becomes intoxicated is subject to the same rules of law as the sober man, it follows : that where a provocation has been received which, if acted upon instantly, would mitigate the offense if committed by a sober man, the question in the case of a drunken man sometime is, whether such provocation was in fact acted upon, and evidence of intoxication may be considered in deciding that question : Buswell on Insanity, § 423 ; State *v.* McCants, 1 Speer, 384.

The New York Penal Code defines precisely this question of responsibility in that State in such cases as follows : "§ 22. Intoxicated persons.—No act committed by a person while in a state of intoxication shall be deemed less criminal by reason of his having been in such condition. But whenever the actual existence of any particular purpose, motive, or intent, is a necessary element to constitute a particular species or degree of crime, the jury may take into consideration the fact that the accused was intoxicated at the time, in determining the purpose, motive, or intent, with which he committed the act."

DELIRIUM TREMENS AND THE LAW.

The rule of law is well established both in Engand and the United States, that insanity produced by delirium tremens is a good defense to a criminal charge. Even if induced by intoxication, the victim is no more punishable for his acts than if the delirium had resulted from causes not under his control : Regina *v.* Davis, 14 Cox C. C., 563 ; Bell on Med. Juris. of Inebriety, 9, and cases there cited ; J. Crisp Poole, Med. Leg. Jour., vol. 8, p. 44 ; U. S. *v.*

McGlue, 1 Curt., 1 ; Wharton's Crim. Law (8th ed.), sec. 48,
People v. Williams, 43 Cal., 344 ; U. S. v. Clarke, 2 Cr. C. C.,
758 ; Lanergan v. People, 50 Barb. (N. Y.), 266 ; s. c., 6
Parker, Cr. R. (N.Y.), 209 ; O'Brien v. People, 48 Barb., 274 ;
State v. Dillahunt, 3 Harr. (Del), 551 ; State v. McGonigal,
5 Harr. (Del.), 510 ; Cluck v. State, 40 Ind., 563 ; Bradley
v. State, 26 Ind., 423 ; O'Herrin v. State, 14 Ind., 420 ;
Dawson v. State, 16 Ind., 428 ; Fisher v. State, 64 Ind., 435 ;
Smith v. Com., 1 Duy (Ky.), 224 ; Roberts v. People, 10
Mich., 401 ; State v. Hundley, 46 Mo., 414 ; State v. Sewell,
3 Jones (N. C.), L., 245 ; Cornwell v. State, Mart & Y.
(Tenn.), 147 ; Carter v. State, 12 Tex., 500 ; Boswell v.
Com., 30 Gratt. (Va.), 860 ; U. S. v. Drew, 5 Mason C. C.,
283.

INDEX.